11.04

Mathematics and Meası

D0519997

Co

r

Cover picture: Detail from a facsimile of the Peutinger Table, a copy of a Roman road map. Rome is in the centre. The original is in the Österreichische Nationalbibliothek, Vienna (see fig. 37).

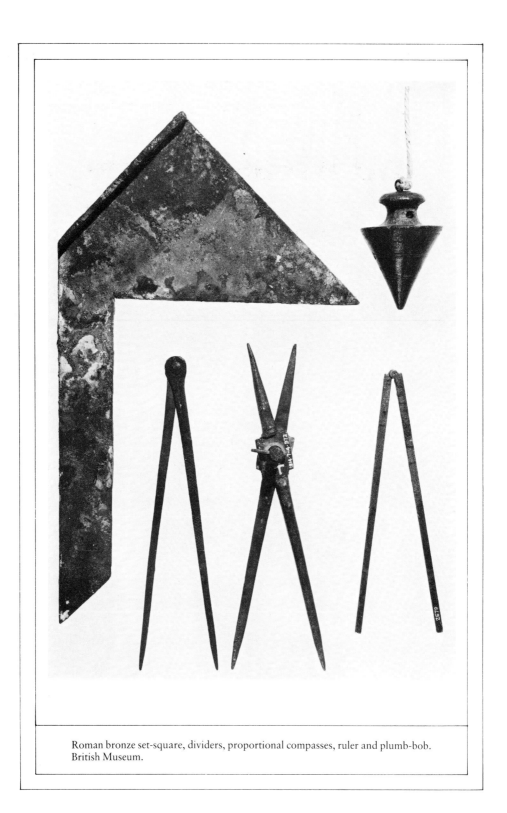

Roman bronze set-square, dividers, proportional compasses, ruler and plumb-bob. British Museum.

Mathematics and Measurement

O. A. W. Dilke

University of California Press/British Museum

Acknowledgements

I am indebted for much help to my wife, a geographer, and our son, a mathematician; to my colleague Dr J. R. Ravetz, Reader in the History and Philosophy of Science, University of Leeds; to Professor B. Wesenberg; to the Museo Egizio, Turin; and to many members of staff of the British Library and the British Museum. These include particularly Dr Ian Carradice, Mr Brian Cook, Mr Ian Jenkins, Mr Thomas Pattie, Dr Jeffrey Spencer, Mr Christopher Walker and Dr Helen Wallis. I wish to thank the publishers for enabling me to reproduce illustrations from my book *The Roman Land Surveyors* (1971), now out of print except in Italian translation; and Thames and Hudson Ltd for facilitating the use of illustrations from my *Greek and Roman Maps* (1985): figs 4 (from J. Ball, *Egypt in the Classical Geographers*, Cairo, 1942), 32, 33, 34 (after Pauly, *Real-Encyclopädie der Classischen Altertumswissenschaft*, supp. x, Stuttgart, 1965), 38, 39.

Thanks are also due to all institutions, acknowledged in the captions, who have kindly supplied photographs, as well as to the following: HBMC (England), fig. 22; Courtauld Institute of Art, London, figs 31, 57 and 58. Artwork by Christine Barratt, Sue Bird, John Dixon and Ann Searight.

Volume 2 in the *Reading The Past* series

Library of Congress
  Cataloging-in-Publication Data

Dilke, Oswald Ashton Wentworth.
  Mathematics and measurement.
  Bibliography: p.
  Includes index.
  1. Mathematics, Ancient.
  2. Mensuration —History.
  I. Title.   II. Series.
QA22.D55 1988   510′.09′01   86–30803
ISBN 0–520–06072–5 (pbk.: alk. paper)

© 1987 The Trustees of the British Museum
Published by British Museum Publications Ltd
46 Bloomsbury Street, London WC1B 3QQ

Designed by Arthur Lockwood
Front cover design by Grahame Dudley

Set in Linotype 202 Sabon and printed at
The Bath Press, Avon

17, 18, 19

Contents

Preface

The sophist Protagoras, who lectured all over Greece in the fifth century BC, proclaimed: 'The measure of everything is human'. Although he spoke in a wide context, his hearers may well have taken him literally. In the Ashmolean Museum, Oxford, is a relief in the shape of a pediment, showing the head, chest, outstretched arms and foot imprint of a man (*below*). Whatever its exact metrological interpretation (and this is disputed), it shows not only awareness of the importance of human measurements, but a sense of harmony and symmetry. Such an approach is visible in Greek and Roman art and architecture, and in the esteem accorded to the teaching of music and geometry.

In our computer age of electronic accuracy, results by ancient methods may seem unreliable. Yet in many respects we can show that the ancients could often come close to an exact measurement, though attempts at modern conversion may be problematic.

The most prominent theoretical aspects of mathematics were astronomy, in Babylonia and Greece, and geometry, in Greece. These sometimes overlapped, and geometry contributed to the development of algebra and trigonometry. Eratosthenes' calculation of the earth's circumference shows a combination of the theoretical and the practical. He may have been more interested in assessing the angle of the sun at midday in different places, but he also based his calculations on a rough approximation to actual land distance, and went on to draw a world map which influenced his successors.

Artefacts from excavations in Egypt, the Athenian agora and elsewhere emphasise the practical importance, for trade and commerce, of standard weights, measures and time-keepers. In the Roman world the practical side of measurement largely predominated. Road-making, surveying, military organisation, water supply and sanitation, all depended on well-defined systems of measuring.

The present work tries to show what a wealth of artefacts we have, particularly in museums and libraries, which throw light on ancient mathematics and measurement, and to set these, alongside technical literature from the Rhind papyrus right down to Renaissance Latin, in their cultural background.

1 Greek metrological relief. Oxford, Ashmolean Museum.

1 The Background

We know that a number of ancient civilisations developed their own techniques of mathematics and measurement separately. Much is now known, for example, of the history of Chinese mathematics and measurement; but they did not influence the knowledge of those subjects in the West, and so it is not proposed to outline them in this book. Of those civilisations which did influence Western developments, the principal ones were Egypt and Mesopotamia, each of which had been evolving its own system from a very early period.

In Egypt we find early interest in astronomy and cosmology, culminating in the division of the year into 12 months of 30 days each, to which five days were added. Egyptian interest in the stars was typified by pictures in tombs showing astronomical features, including maps of star positions, to calculate the passage of night hours. Such information may have been of practical help to a farming community such as flourished in the Nile valley. The Egyptians went on from this to devise a system of 12 daytime hours and 12 night hours, which formed the origin of the modern system of division of the day. Since, however, the amount of daylight varied according to the month, it will be seen (chapter 7) that some careful observations and calibrations were involved.

Egypt is the country of the Nile and the pyramids, and it is not surprising that great accuracy was attempted with regard to both of these. The annual autumn flooding of the Nile, caused as we now know by high summer rainfall in Ethiopia, helped the fertility of Egyptian fields by spreading quantities of silt. To record levels upstream, a Nilometer was set up on Philae Island, which is now submerged by the lake of the Aswan dam. Strabo describes it as a well, in which the level of the Nile could be read from marks on the side. From such readings, scribes could register the maximum, minimum and mean levels of the water in cubits. As is known from ancient writers, annual compilations of such readings were carefully preserved, but these have not themselves survived. A Nilometer at Memphis (Cairo) is mentioned by Heliodorus, and similar gauges have been found at Edfu, Luxor and elsewhere, with cubit markings mostly between 52.8 and 53.3 cm above the river's low level.

When the floods receded, many landowners' boundary marks had inevitably been washed away, so it was important that surveying should be carried out immediately. There are Egyptian representations of surveyors employing knotted ropes (the knots indicating subdivisions of linear measurement), the *merkhet* (a split centre-rib of a palm-leaf, used for

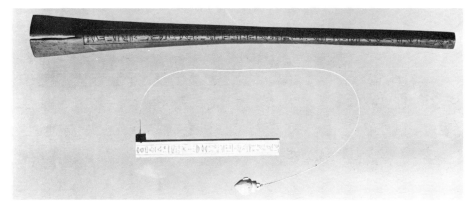

2 Reproduction of a *merkhet*, an Egyptian surveying instrument. The *merkhet* was aligned on an object by looking through the split centre, held upwards. It was used with a short plumb-line and plummet. London, Science Museum.

sighting), and measuring rods. The priests inaugurated this rapid re-survey of the land, which had to be ready for winter cultivation. There is no evidence that land survey maps were used in dynastic Egypt for this operation. But by means of exact area measurements and verbal descriptions, the status quo was re-established. Graeco-Roman writers from Herodotus (*c.* 484–*c.*420 BC) right down to Cassiodorus (*c.* AD 490–*c.* 583) attribute the origins of geometry, literally 'measuring of the earth', to this practice.

Egyptian numerals were based on the decimal system, the highest values being placed on the left. Up to nine of the same symbol could be used, arranged in one or two lines. The hieroglyphic signs are:

1	ǀ	10,000	𓆼
10	∩	100,000	𓆐
100	૭	1,000,000	𓁨
1000	𓎆		

Examples are:

465   𓍢∩∩∩∩∩ǀǀǀǀǀ

4323   𓆼𓆼𓆼𓆼𓍢𓍢𓍢∩∩

The sign for 1,000,000 disappeared in early times, and after its disappearance a system of multiplication, by placing one number over another, was sometimes used; for example,

$$1,100,000 = 100,000 \times 11 = \text{𓆐}_{∩ǀ}$$

The most famous mathematical work from dynastic Egypt is the Rhind mathematical papyrus, copied by the scribe Ahmes or Ahmose from a papyrus of 1849–1801 BC. It was bought by A. H. Rhind at Luxor in 1858 and is now in the British Museum. In its geometry we can see some understanding of the properties of right-angled triangles, but those with shorter sides 4:3 were chosen to illustrate pyramidal ratios (p. 9) rather than Pythagoras' theorem. Also included is an approximation for π, the ratio of the circumference of a circle to its diameter; the value given is $(\frac{16}{9})^2 = 3.1604938$, too large by about 0.019.

3 Part of the Rhind mathematical papyrus, dealing with the calculation of areas. Second Intermediate Period, *c.* 1575 BC. British Museum.

The Rhind papyrus has several arithmetical problems, such as the following: 'A quantity whose half is added to it becomes 16'; in other words, 'find $\frac{2}{3}$ of 16'. The workings are as follows:

<table>
<tr><td colspan="2">(a)</td><td></td><td colspan="3">(b)</td></tr>
<tr><td>1</td><td>2</td><td></td><td>*</td><td>1</td><td>3</td></tr>
<tr><td>$\frac{1}{2}$</td><td>1</td><td></td><td></td><td>2</td><td>6</td></tr>
<tr><td></td><td></td><td></td><td>*</td><td>4</td><td>12</td></tr>
<tr><td></td><td></td><td></td><td></td><td>$\frac{2}{3}$</td><td>2</td></tr>
<tr><td></td><td></td><td></td><td>*</td><td>$\frac{1}{3}$</td><td>1</td></tr>
</table>

Table (a) shows that the divisor should be 3. The purpose of table (b) is to find one-third of 16. Column 2 multiplies column 1 by three. The asterisked items are added together because the corresponding numbers in the right-hand column add up to 16. This establishes that $1 + 4 + \frac{1}{3}$—that is, $5\frac{1}{3}$—is one-third of 16. Since $\frac{2}{3}$ of 16 is to be found, the remaining workings are:

<table>
<tr><td colspan="2">(c)</td></tr>
<tr><td>1</td><td>$5\frac{1}{3}$</td></tr>
<tr><td>2</td><td>$10\frac{2}{3}$</td></tr>
</table>

Geometry was equally required for the construction of pyramids. In the first place, care was often taken to achieve orientation north, south, east and west. North–south orientation could easily be obtained by finding the direction of the noonday sun. A vertical pole was set up in the sand as a gnomon. Then the path traversed by the tip of its shadow could be observed. The points A and B, where this intersected a suitable circle drawn around the gnomon, would be joined. Then the line AB was bisected to establish the direction of the sun at mid-day. A possible alternative method, outlined by I. E. S. Edwards in *The Pyramids of Egypt*, would have been to bisect the angle formed by the rising and setting positions of a star.

Egyptian pyramids (with the exception of the most ancient, the Saqqara step pyramid) were square in ground-plan and fully pyramidal in shape. The specification for the gradient, known as the 'batter', of such pyramids was denoted by the hieroglyphic word *śkd*, meaning ratio. It was expressed in terms of the number of palms, in half the length of a side, per cubit of vertical height (1 cubit = 7 palms). The Rhind papyrus is particularly concerned with the geometry of pyramids, measurements being reckoned in royal cubits (see chapter 4). An example of finding the batter is: 'A pyramid whose vertical height is $93\frac{1}{3}$ [cubits]. Let me know its batter, 140 [cubits] being the length of its side.' Half the length of a side is 7×70 palms; the batter is therefore:

$$7 \times \frac{70}{93\frac{1}{3}} = 7 \times \frac{210}{280} = 5\frac{1}{4} \text{ palms;}$$

since 1 palm = 4 finger's-breadths, this is expressed as 5 palms, 1 finger's-breadth.

The dimensions and orientation of the Great Pyramid were very carefully fixed. Like a number of dynastic Egyptian buildings, its sides face the four points of the compass. Measurements are in whole numbers of royal cubits, often rounded off, which vary only narrowly—mostly between 52.3 cm and 52.5 cm, with an average of 52.36 cm. The mean of the four base lengths, which are very nearly equal, is 440 cubits. The original height from platform to apex was 280 cubits, giving a batter, as defined above, of $5\frac{1}{2}$ palms. The length of the descending passage is 75 cubits. The floor length of the grand gallery is 88 cubits, one-fifth of the base length. The king's chamber is 20 cubits long and 10 cubits broad; its height is 11 cubits, again a factor of the base length, at 52.5 cm to the cubit.

Contrasting somewhat with this precision is the clumsiness of the Egyptian method of expressing fractions. Apart from $\frac{2}{3}$, only 'simple' fractions were used, that is, those having the numerator 1. Thus we find statements like: '$\frac{1}{11}$, $\frac{2}{3}$ of it is $\frac{1}{22} + \frac{1}{66}$'. Fractions such as $\frac{2}{33}$ were not used.

In linear measurement it was customary, as we have seen, to divide the palm into finger's-breadths. These were similar in length to the palms and finger's-breadths of the Greek Olympic foot. But, although the Greeks borrowed geometrical technique from Egypt, it is not likely that they took units of length, which can always be coincidentally similar when based on the human body.

Some specimens survive of maps which were obviously based on measurements of the region depicted, though they do not include figures. We may mention the Turin papyrus and a number of temple or garden plans. The Turin papyrus (fig. 4), now in the Museo Egizio, Turin, but incomplete, dates from *c.* 1300 BC and depicts a mining area near Umm Fawakhir, between the Nile and the Red Sea. This is a very early example in the history of cartography of the use of colour to differentiate between various types of mineral and rock. No measurements are given, but since the map is thought to have been drawn up in connection with a legal dispute in a topographically identifiable location, we must imagine that some scale was implicit in it. The large-scale plans likewise record no measurements, and sometimes there is a mixture of plan and elevation; but temple or garden measurements were easily obtainable and could without difficulty be represented on an agreed scale.

An equal concern for accuracy is found in the Mesopotamian civilisations. The countries of the Tigris and Euphrates had much foreign trade with other areas, exchanging items such as spices, jewels and silks from the East for minerals or timber from the West. This international trade was dependent on recognised units of weight, volume, area and length. The concept of money may have arisen from the Mesopotamians' use of metal bars with a mark indicating their weight.

Mesopotamian writing, known as cuneiform from the Latin *cuneus*, 'wedge', was executed on clay tablets. Symbols were built up from wedge shapes formed with a square-ended stylus. However, in early Sumerian writing (*c.* 3000 BC) the numerals were incised with a different instrument, with rounded ends, one large, one small. It may have been a reed or a wooden spike. The Sumerians used the sexagesimal system, which had a base of 60, but there were individual symbols for 36,000, 3600 ($=60^2$), 600, 60 and 10. In late Sumerian (*c.* 2500–2000 BC), numerals, like other written symbols, were made with a square-ended

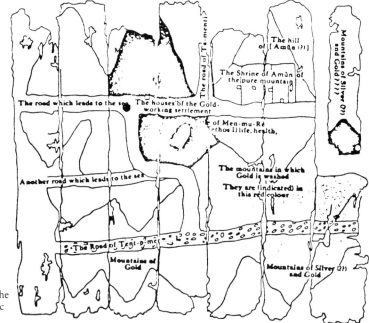

4 The main fragments of the Turin papyrus (hieroglyphic text translated by J. Ball)

stylus. Owing to the difficulty of making curved lines with this, rhomboids took the place of circles.

In the later Babylonian script (c. 2000 BC–c. AD 75) there is a different system for recording numbers. We find two notations: the old sexagesimal system, still used for mathematics and astronomy, and also the decimal system, used for everyday purposes such as trading and accounting. A complication arises in this later script, namely that in its sexagesimal notation only two signs in all were used: one for 216,000, 3600, 60 and 1, the other for 36,000, 600 and 10. A slight break between two signs indicated a change in the power of 60.

SUMERIAN AND BABYLONIAN NUMERALS

Early Sumerian (c. 3000 BC)

⊙ = 60 × 60 × 10   ○ = 60 × 60   ◉ = 60 × 10   ▽ = 60   ○ = 10   ᵥ = 1

So

oᵥ = 11   ▽ᵥ = 61   ▽oᵥ = 71   ◉▽oᵥ = 671   ○◉▽oᵥ = 3600 + 671 = 4271

Later Sumerian (c. 2500–2000 BC)

◈ = 60 × 60 × 10   ◇ = 60 × 60   ✳ = 60 × 10   Ꭲ = 60   ◂ = 10   Ꭲ = 1

So

◂Ꭲ = 11   ꓔ Ꭲ = 61   Ꭲ◂Ꭲ = 71   ✳Ꭲ◂Ꭲ = 671   ◇✳Ꭲ◂Ꭲ = 3600 + 671 = 4271

Babylonian (2000 BC–AD 75)

1 = Ꭲ   10 = ◂   60 = Ꭲ   60 × 10 = ◂   60 × 60 = Ꭲ   60 × 60 × 10 = ◂

and so on. Everything depended on the order of the wedges, reading from left to right. So ◂Ꭲ = 11. But it can also equal 601 (60 × 10 + 1). ꓔꓔ = 2, but Ꭲ Ꭲ = 61; the difference depends only on careful writing and reading.

$$100 = Ꭲ \text{⁂} \quad (60 + [4 \times 10]) \qquad 1000 = ◂Ꭲꓔꓔ\text{⁂} \quad (16 \times 60 + [4 \times 10])$$

Both the linear measurements and the capacity measurements (chapter 4) of the Meso-potamians were to some extent linked with the sexagesimal number system mentioned above. The cord, the largest linear unit used by surveyors, represented 60 double cubits. The double cubit was divided into 60 digits (finger's-breadths), the smallest linear unit. A tablet containing a plan of Nippur, c. 1000 BC, gives measurements for buildings which seem to be in units of 12 cubits (about 6 m). An earlier plan, c. 2300 BC, from Yorghan Tepe near Kirkuk, mentions the area of a plot of land shown, 354 *iku* (about 125 ha).

For capacity measurement the largest unit, the *gur*, represented 300 *sila*, the smallest. But although we can give approximate equivalents, modern attempts to correlate Sumerian weights with Sumerian capacity measurements have been inconclusive.

In mathematics we may think that the Babylonians were more proficient in arithmetic, the Egyptians in geometry. Thus, to take an example where arithmetic counts for more than geometry, a tablet in the Yale Babylonian collection shows a square with sides 30. On the diagonal are two sets of figures: 1, 24, 51, 10 and 42, 25, 35, which correspond to $\sqrt{2}$ and $30 \times \sqrt{2}$ respectively. This means that the Babylonians had a very close approximation to $\sqrt{2}$, since 1, 24, 51, 10 equals

$$1 + \frac{24}{60} + \frac{51}{3600} + \frac{10}{216,000}$$

which is 1.41421296: the true value is approximately 1.41421356. For π, on the other hand, the Babylonians tended to use 3, while the Egyptians certainly used as an approximation $(\frac{16}{9})^2$ (see p. 8). They may also have worked with $\pi = 3\frac{1}{7}$ (3.$\dot{1}$4285$\dot{7}$), the true value being 3.14159 . . . .

The Babylonians had extensive lists of squares, which they probably used to aid multiplication. Similarly, there are lists of reciprocals expressed in the sexagesimal system:

2	30	6	10
3	20	8	7, 30
4	15	9	6, 40
5	12		

That is to say, $\frac{1}{4} = \frac{15}{60}$, while $\frac{1}{8} = \frac{7}{60} + \frac{30}{3600}$, and so on. If necessary, further subdivisions were introduced, e.g.:

$$27 \qquad 2, 13, 20$$

The principal application of Babylonian mathematics was to astronomy, in which many achievements are recorded. As Ptolemy noted, almost complete records of eclipses were preserved from $c.747$ BC. The astronomers established that solar eclipses can only take place at the end of a lunar month, and lunar eclipses only in the middle of one. Whether they were able to predict eclipses before about 300 BC is uncertain. By 700 BC at the latest, and almost certainly earlier, they had named three regions of the sky after gods. The region of Enlil was to the north; that of Anu, covering about 34° in declination, was equatorial; and that of Ea lay to the south. But there is also a group of early texts which divides the sky into three zones, each containing 12 sectors. These zones bear names of planets and constellations with progressive numbers. Whereas in early times observation of Venus, for example, may have been connected with divination, by about 400 BC a body of mathematical astronomy had been built up, using a zodiac with 12 signs, each of 30°. This constituted a system of reference for the movements of the sun and planets which was passed down to the Alexandrian Greeks and their successors.

In the fifth century BC the 19-year cycle known as the Metonic cycle (see chapter 3) was incorporated in the Babylonian calendar; this is now thought to have come before Meton unsuccessfully proposed its incorporation in the Athenian calendar.

Halley's comet, which reappears roughly every 75 years, was observed in detail both by the Chinese and by the Babylonians. The Babylonian record dates in all probability to 164 BC, and reads in translation: 'The *sallammu* [comet] which previously appeared to the east in the path of Anu in the area of the Pleiades and Taurus, to the west [. . .] and passed on in the path of Ea . . .' The regions of the sky are the same as those given above.

Among other countries which had some influence on the mathematical progress of parts of the Graeco-Roman world were Minoan Crete, Phoenicia and Etruria. Minoan Linear B, the pictographic and syllabic script which influenced Cypriot writing, had a system of numerals somewhat similar to the Egyptian. The Phoenicians used a knowledge of astronomy drawn from Babylonia to help with navigation by night; and the Greek alphabet, used for one of the two Greek systems of numeration (chapter 2), was derived from the Phoenician. The Etruscans used methods of divination, including observation of quarters of the heavens, which seem to have had a connection with Babylonian practice. Like the Babylonians, they named these areas of observation after deities. Early Greek scientists went from Asia Minor to Egypt in the sixth century BC and consulted the priests. Later Greeks had access to Babylonian knowledge through the writings of Berosus, who flourished about 290 BC. It is significant that throughout the Graeco-Roman period these aspects of Egyptian and Babylonian civilisation stood in high esteem; though after a time Greek mathematicians considered they had advanced beyond their Near Eastern predecessors.

2 Numbering by Letters

The Greeks and the Romans had systems of numbering which, compared with arabic numerals, seem awkward to us. It is recognised today that most of the Minoan–Mycenaean script known as Linear B was in a language which may be considered the ancestor of Greek. This script (*c.* 1450–*c.* 1200 BC) had a system of numeral abbreviations as follows:

1	I	5	III / II
10	–	50	☰☰
100	◦		
1000	✧	500	◦◦ / ◦
10,000	✦	5000	✧✧ / ✧

These abbreviations, among the first to be deciphered, could be combined with ideograms, such as ⚲ for 'woman'. But the script disappeared in the dark age which followed. The only possible relic of Linear B—which, however, is paralleled in other systems of numeration that are definitely independent—is the series I, II, III, IIII adopted for 1–4 in the 'acrophonic' system of classical Athens (p. 14).

From about the sixth century BC the Greeks and later the Romans had methods of expressing numbers which were based largely or wholly on the alphabet. Where many numbers were involved, any system of abbreviation was an improvement on the early practice of writing them out in full. But Greek city states, being independent, frequently did not collaborate. So we are not surprised to find two different systems emerging, one at Miletus in Asia Minor, the other at Athens and in Attica. It is now thought that the Milesian numeration started first: it is to be found on early Attic vases, though from about 480 BC both systems are found.

The Milesian or alphabetic numeration may possibly have been invented by one of the two earliest Greek scientists, Thales (who flourished in 585 BC) and Anaximander (*c.* 610–540 BC), both of whom lived in Miletus. As an example of the applied mathematics of the time, Anaximander is said to have designed the first Greek map. If he put distances on it, he would certainly have needed a set of abbreviations.

About the mid-eighth century BC the Greek alphabet, which later strongly influenced the Roman (and much later still the Cyrillic), had been adapted from the Phoenician. Successive letters were used to denote numbers, though they had no connection with the pronunciation of the numerals. As they have come down to us, they are usually minuscule (lower-case) letters, except on inscriptions. Thus 1 was α (alpha), 2 was β (beta) and so on to 10, which was ι (iota). For the multiples of 10 up to 90 and of 100 up to 900, successive single letters to the end of the alphabet were used. Thus, 111, for example, was expressed by the letters standing for 100, 10 and 1. Early usage in Asia Minor was in the reverse order. Since 27 letters of the alphabet thus needed to be used for denoting numbers up to 999, not only the current alphabet but three obsolete or obsolescent letters were used.

∢	A	Alpha
٩	B	Beta
𐤂	Γ	Gamma
△	△	Delta
⋛	E	Epsilon

5 The first five letters of the Phoenician and Greek alphabets compared.

Numerals from 1000 upwards started again from α, but with a stroke written below. We may tabulate the Milesian numeration thus:

1	α	10	ι	100	ρ	1000	α
2	β	20	κ	200	σ	10,000	M
3	γ	30	λ	300	τ	20,000	β͵M
4	δ	40	μ	400	υ		
5	ε	50	ν	500	φ		*Examples*
6	F	60	ξ	600	χ	11	ια
7	ζ	70	ο	700	ψ	63	ξγ
8	η	80	π	800	ω	128	ρκη
9	θ	90	ϙ	900	⅀	1601	αχα

The early letters brought in were wau (6), a form of digamma; koppa (90); and sampi (900). M stood for *myrioi*, 10,000, and a numeral could be placed over it to denote compounds.

These abbreviations became so standard that the full system lasted right down to the fall of Constantinople (1453), and they are used to a limited extent in modern Greek, though they were never adapted to languages other than Greek. They had serious disadvantages, one being the mastering of 27 symbols, though the ancients were more accustomed than we are to memorising such things. But also, as in the other Graeco-Roman numerations, a number like 222, for example, is not represented by three identical signs. In this case, rather, it uses the 20th, 11th and 2nd letters respectively of the enlarged Greek alphabet.

There were two methods of expressing fractions in the Milesian numeration. First, as in Egyptian notation, the Greeks often represented them as the sum of unit fractions: e.g., $\frac{2}{5} = \frac{1}{3} + \frac{1}{15}$, which was expressed by letters of the alphabet with accents—in this case γ′ιε′. (There was no plus sign in Greek mathematics; numbers to be added together were put side by side, with or without *kai*, 'and'.) As special signs we find $L = \frac{1}{2}$, $β' = \frac{2}{3}$ (not $\frac{1}{3}$ as might be expected). Secondly, in mathematical writings we also find non-unit fractions, with numerator and denominator written either on the same line or on two lines. The Milesian system was adapted by several mathematicians to express enormous numbers. Archimedes in his *Sand-reckoner* devised a special notation (not used in practice) which could express numbers up to what in our own notation would be 1 followed by 80,000 million millions of zeros.

The other Greek numeration system is usually known as Attic or acrophonic. 'Attic' means that it was mainly used in Athens and the rest of Attica; 'acrophonic' that it was based (apart from the symbol ǀ for 1) on the first sound of the numeral. Another name is 'Herodianic', from the grammarian Herodian who described it in the second century AD. This numeration used capital letters, and 100 was represented by H, standing for the rough breathing, indicating aspiration, at the beginning of the word *hekaton*, an abbreviation particularly common in some Aegean islands. These are the normal forms:

1	ǀ	500	Ͱ		*Examples*
5	Γ	*(pente)*	1000	X *(khilioi)*	11 Δǀ
10	Δ	*(deka)*	5000	Ͷ	63 ͰΔǀǀǀ
50	Ͱ, Ͳ, Ͷ		10,000	M *(myrioi)*	128 HΔΔΓǀǀǀ
100	H	*(hekaton)*	50,000	Ͷ	1601 XͰHI

It will be seen that these abbreviations, which somewhat resemble Roman numerals, were more self-explanatory, though lengthier, than the Milesian. Fractions were uncommon: C stands for $\frac{1}{2}$, while Ɔ seems intended for $\frac{1}{4}$.

This Attic numeration did not last as long as the Milesian, fading out in the third century BC and disappearing under the Roman Empire. An interesting feature, visible in the table above, is that 50, 500, 5000 and 50,000 were formed by monograms incorporating the symbols for 5 and, respectively, 10, 100, 1000 and 10,000. Like the Roman numeration, however, repetition of letters was used to form some numerals: e.g., as Δ stood for 10 (*deka*), ΔΔ stood for 20, though 20 was written in full as *eikosi* (εἴκοσι) and so pronounced— not, of course, as *deka deka*. Because of its use of capital letters, this numeration was particularly suitable for inscriptions. In fact the forms used are all ones that can be expressed by straight lines, which facilitated carving on stone.

In the Greek world there was no sign for zero in regular use, but a sign O or $\overline{O}$ is found in astronomical writings, for example in the tables of solar and lunar eclipses in Book VI of Ptolemy's *Almagest*.

The Roman numeration is far more familiar to us, since it is still widely used: occasionally for year-dates, frequently for numbering prefatory pages in books, and so on. Many have supposed that M and C originated as initial letters of *mille* ('thousand') and *centum* ('hundred'). However, the other letters do not fit such a pattern, and indeed none originated in that way. The early Latin abbreviation for 1000 was ⓪, later CIↃ, and it was the second half of this, IↃ or D, which came to be used for 500 (*quingenti*). That symbol ⓪ had its origin in phi (φ), which formed part of the alphabet in the Chalcidian Greek colonies of southern Italy, founded from Chalcis in Euboea. The Latin alphabet was based on the Chalcidian Greek alphabet, a variant of which (with E and F, digamma, transposed) appears on the Formello vase found near Veii:

6 The Greek alphabet on a vase from Formello, north of Rome. (After delta comes digamma; after nu, Phoenician samekh; after pi, Phoenician shin; then koppa; after upsilon come xi, phi and chi.)

Parallels with Etruscan have suggested that three letters of this alphabet that were not used in the Latin alphabet, the ninth and the last two, came into use as numerals, after slight changes: as C for 100, probably influenced by *centum*; ∞ for 1000; and ⊥ for 50. From the abbreviation M.P. for *mille passus* (mile) or *milia passuum* (miles), M came into use for 1000, being in its written forms not unlike the previous symbol. Then ⊥, though not influenced by an initial letter, developed into the closest letter in appearance, L.

Since the scheme was clearly derived from the Chalcidian alphabet, why can we not conjecture that after the first symbol, a stroke for 1, the next two symbols, V and X, originated in the last two Chalcidian letters? For, rather than the forms found on the Formello vase, the regular forms for υ and ξ were actually V and X. Instead, attempts are made to explain V as half of X or X as a double V. Such an origin is admittedly like that of D for 500, but does not explain why 50 and 100 were not similarly treated.

The subtractive forms, such as IV, which we now use where a number is one short of a multiple of 5, represent part of an alternative method used by the Romans. C. M. Taisbak has shown that this method was extremely well suited to the abacus. Since 28 could be either *viginti octo* ('twenty eight') or *duodetriginta* ('two from thirty'), it came to be abbreviated as either XXVIII or, less commonly, XXIIX. Although 4, 14, etc., had no such alternative verbal forms, they could be expressed by analogy as IV, XIV, etc., as well as in the commoner forms IIII, XIIII, and so on. Under the late Republic, lines came to be used to denote thousands, thus: $\overline{VI}$ = 6000; $\overline{|VI|}$ = 600,000. For ½, the letter S (*semis*, whence

English 'semi-') was adopted. For other fractions the duodecimal system was used, so that from one to five dashes, representing the number of twelfths, were used, preceded by S or not as appropriate. Thus $\frac{1}{4}$ was expressed as =- $(\frac{3}{12})$, $\frac{1}{3}$ as == $(\frac{4}{12})$, $\frac{2}{3}$ as S = $(\frac{1}{2} + \frac{2}{12})$.

The integral number abbreviations, with alternative forms where applicable, may be tabulated thus:

1	I, I	500	Ɔ, Ә, D	
5	V	1000	Ⅽ, ⅭƆ, ∞, later M	
10	X	5000	ⅭƆƆ ; also V̄, Ә	
50	⊥, ⊥, L	10,000	ⅭⅭƆƆ ; also X̄, Ⓓ	
100	Ⅽ			

Thus 3100 could be expressed as ⅭƆ ⅭƆ ⅭƆ Ⅽ or as I̅I̅I̅C, later also as MMMC.

Some additive and subtractive forms

4	IIII, IV	40	XXXX, XL
8	VIII, IIX	400	CCCC, CD
9	VIIII, IX		

As an example of the use of Roman numerals in Latin, this is an extract from the *Corpus Agrimensorum* explaining how to calculate the volume, in (cubic) feet, of a rectangular cistern: *Si fuerit arca longa ped. XXX, lata ped. XV, alta ped. VII, duco longitudinem per altitudinem: fiunt ped. CCX. hoc duco per latitudinem: fiunt ped. I̅I̅ICL.* 'If a cistern is 30 ft long, 15 ft wide and 7 ft high, I multiply the length by the height: this makes 210 [sq.] ft. I multiply that by the width: this makes 3150 [cu.] ft.'

The question is often asked today how difficult the Greeks and Romans found it to add, subtract, multiply and divide. Materials for assessing this are rather lacking, but it is thought that there were few problems, except that long division was awkward by classical methods. For addition and subtraction the abacus must have speeded things up. For multiplication, tables such as that in fig. 7 could be used. The examples of these that have come down to us are more commonly in Greek numerals, as writing materials have survived better in the sands of Egypt or in other dry parts of the eastern Roman Empire, where Greek was the *lingua franca*.

7 Greek multiplication table on a wax tablet. Lines 2–4 may be rendered 2 × 1 = 2; 2 × 2 = 4; 2 × 3 = 6. British Library, Add. MS 34186 (1).

3 Mathematical Education in the Greek World

To the Greeks the teaching of mathematics was an extremely important aspect of education, while to most Romans it was viewed as a necessary aid to technology. It has to be admitted that Greek and Roman writers do not tell us all we want to know about the teaching of mathematics, of which measurement was an inherent part. The word *mathēma* itself, a noun from the verb *manthanō*, means 'subject connected with learning', which shows how basic a concept it was in Greek education. Its two main branches in antiquity, arithmetic and geometry, mean respectively 'subject of numbers' (*arithmos* = number) and 'measuring (*metrein*) of the earth (*gē*)'. The word *geōmetria* came from the use of geometry in surveying and in general denoted either surveying or geometry. It was perhaps to avoid this ambiguity that Euclid called his manual in 13 books not *Geōmetria* but *Stoicheia*, 'elements'. The name 'algebra' is Arabic, though not only had some of the principles of algebra been discovered by the Babylonians, but the Greek mathematician Diophantus developed it in some depth.

Geometry was above all the type of mathematics admired and exalted by the Greeks. To them it seemed even more than arithmetic to point to perfection. We are told by Vitruvius that the philosopher Aristippus of Cyrene, who flourished *c.* 400 BC, was shipwrecked on an island with some companions. Aristippus, who liked the pleasures of life, spotted geometrical drawings recently executed in the sand, and reassured his companions by saying, 'All is well: I see traces of mankind.'

The philosophers of the Ionian school, from Miletus and elsewhere in Asia Minor, were the first (sixth and fifth centuries BC) to use mathematics extensively in the Greek areas. They asked questions about the world and the universe, and tried to give answers to these questions. The story in Herodotus that Thales, founder of the school, predicted a total eclipse of the sun in western Asia Minor is doubted by modern scholars. But Thales or his successors do seem to have applied geometry far more widely than the Egyptians had. From this time onward it became in many ways the foremost academic discipline. Four propositions ascribed in antiquity to Thales (if correctly, he is perhaps unlikely to have thought of them as propositions) are:

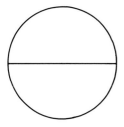

1. A circle is bisected by its diameter.

2. The angles at the base of an isosceles triangle are equal (cf. Euclid i.5).

3. Two intersecting straight lines form two pairs of equal angles.

4. An angle inscribed in a semicircle is a right angle.

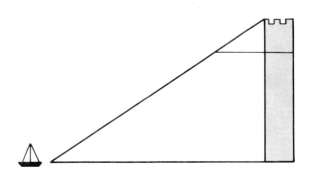

He is also said to have worked out the distance of a ship from a high point on shore by the use of similar triangles (above).

About 531 BC Pythagoras of Samos emigrated to Croton in southern Italy and set up a community of mathematicians and mystics. Though best known today for the 'theorem of Pythagoras', his greatest achievement perhaps lay in establishing the mathematical basis of musical intervals. What we know as 'Pythagoras' theorem'—that the square on the hypotenuse of a right-angled triangle is equal to the sum of the squares on its other two sides—embodies a fact which was known to Babylonian mathematicians from very early times. The Greeks ascribed the proof of the fact to Pythagoras, and this is very probably correct, but if so we do not know how he demonstrated it. Probably the proof was what we may call the Chinese puzzle type, of which one version is known from Chinese mathematics, rather than by the Euclidean method (fig. 8). This latter was taken up by Arabic textbooks and became the standard approach in our geometry manuals of the nineteenth and early twentieth centuries.

For a fragment explaining the application of 'Pythagoras' theorem' to the measurement of a trapezium we may turn to a Greek papyrus at the Field Museum of Natural History, Chicago, described by E. J. Goodspeed. The figure accompanying it, as reconstructed (fig. 9), gives all linear and area measurements. But it is clear that originally only the exterior linear measurements were given; for the base we are in effect only told that the total is 16, while the component parts are what we should express algebraically as $x + 2 + x + y$. These data, however, enable first the remaining linear measurements (5, 12, 4) to be found and then the total area, $30 + 24 + 30 + 24 = 108$.

Among other achievements may be mentioned that of the Athenian astronomer Meton (c. 435 BC), who aimed at synthetising the solar year of $365\frac{1}{4}$ days and the lunar year of $354\frac{1}{3}$ days. He determined that 235 synodic (lunar) months are approximately equal to 19 years, to such effect that our calculation of the date of Easter, using the Golden Numbers I–XIX, depends ultimately on his cycle, improved by later Greek astronomers and reinforced

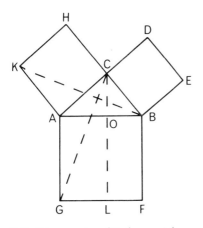

8 Euclidean version of 'Pythagoras' theorem' (Euclid, i.47).

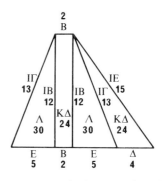

9 Reconstruction of a diagram in the Field Museum Papyrus 1, Chicago: the application of 'Pythagoras' theorem' to the measurement of a trapezium.

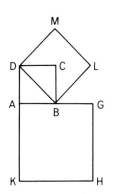

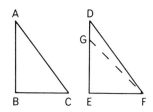

10 Diagram illustrating a problem in Plato's *Meno*: doubling the square.

11 Euclid i.26: to prove that two triangles are equal in all respects if they have one side and two angles each equal. This figure is for the case where the side is between the angles. Here BC = EF, ∠ ABC = ∠ DEF, ∠ ACB = ∠ DFE. If on DE we cut off EG = BA, then, from Euclid i.4, triangles ABC, GEF are equal in all aspects. So ∠ GFE = ∠ ACB. But ∠ ACB = ∠ DFE, so unless G coincides with D we are making the part equal to the whole.

by borrowings from the Near East. Whether he was indebted to Babylonian astronomers is disputed.

With this background, on which Greek mathematicians were constantly building, we need not be surprised that mathematics featured high in the educational curriculum. By the late fifth century BC a theory of mathematical education had been formulated. Much of the teaching at the higher level was carried out by sophists who travelled from city to city throughout the Greek world. The sophist Hippias of Elis (*c.* 485–415 BC), who boasted of the wealth he acquired from such teaching, advocated a curriculum consisting of arithmetic, geometry, astronomy, and acoustics.

Plato (*c.* 429–347 BC) was a great advocate of geometry, and in his late work *The Laws* he supported Hippias' curriculum. The inscription on Plato's door, or that of his Academy, ran ΑΓΕΩΜΕΤΡΗΤΟΣ ΜΗΔΕΙΣ ΕΙΣΙΤΩ, literally 'no one ungeometrical may enter', i.e., 'only for the geometrically literate'. According to him, his master Socrates, by means of leading questions, used in his dialogue *Meno* two geometrical problems involving measurement to support his theory that we can all recollect something from a former life. In the first, he showed that twice the square on a straight line is not the square on twice that line. In this diagram (fig. 10) the given square is that on AB, while the double square is not AGHK but BDML, the square on the diagonal. But Plato did not regard geometry as valuable because it had a practical end. To his mind it was to be studied as contributing towards the ideal life aspired to by a philosopher.

The oldest extant manual of geometry is that by Euclid of Alexandria (*c.* 300 BC), which built on earlier works by Hippocrates of Chios (*c.* 470–400 BC), Eudoxus of Cnidos (*c.* 390–*c.* 340 BC) and others. Euclid's textbook, of which Books 1–6 dealt with plane geometry, remained a standard work in schools until quite recent times. Books 7–9 deal with the theory of numbers, Book 10 with irrationals and Books 11–13 with solid geometry. An example of a theorem affecting measurement is i.26 (fig. 11), taken up with i.15 by Roman surveyors (fig. 12) as a means of measuring a river without crossing it. Euclid was a great believer in methodical progress from one deduction to another, and in not taking anything for granted.

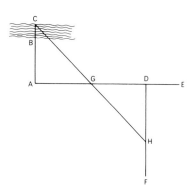

12 M. Iunius Nipsus in the *Corpus Agrimensorum*: to measure the width BC of a river without crossing it. Produce CB to A, draw AE at right angles. Mark off any point D on AE, and draw DF at right angles to AE. Bisect AD at G, produce CG to meet DF at H. Then it can be proved that DH = AC, and by subtracting AB the length of BC is found. Reconstruction by M. Cantor (1875), simplified.

The most gifted mathematician of antiquity was Archimedes of Syracuse (*c.* 287–212 BC), killed by a Roman soldier at the siege of his city. His tomb represented a cylinder circumscribing a sphere, with the ratio of their respective volumes 3:2, which he discovered. He upheld the use of approximation where he felt that certainty could not be attained. Thus, to reach an approximation for π, he worked out the approximate circumferences of regular polygons of 96 sides, one inscribed in a circle and one circumscribing it. The ratios of those to the diameter of the circle were $3\frac{1}{7}$ and $3\frac{10}{71}$. In decimal form these are approximately 3.14286 and 3.14084, compared with the correct value (to five decimal places) 3.14159. He also gave approximations to square roots. In order to discover theorems relating to the surface area and volume of a sphere, the volume of a conoid and the area of a figure bounded by a parabola and a straight line, he imagined the weighing of figures to obtain ratios.

The most famous story is given by Vitruvius in his preface to Book 9 of *De Architectura*. King Hiero had been told that a certain amount of silver had been substituted for an equal weight of gold in a crown made for him:

> When Archimedes was investigating this, he happened to visit the baths, and as he was going down into the bath-tub, he noticed that the displacement of water was equal to the immersed part of his body. This suggested to him a means of solving the problem. Without delay, in his delight he leapt out of the bath, and making his way naked towards his home he indicated in a loud voice that he had found the answer. As he ran, he repeatedly shouted '*heureka, heureka*'.

As the text goes on to explain, he proved that a mass of gold the same weight as the crown displaced less water than a mass of silver of the same weight, since the denser metal (gold) of equal weight will occupy the lesser volume.

The contributions to measurement made by Eratosthenes and Hero of Alexandria are largely connected with map-making and surveying, and as such are discussed in chapters 5 and 6. Applied arithmetic was not covered in manuals bearing the title of arithmetic, such as that, preserved in excerpts, of Nicomachus of Gerasa (second century AD). To such writers arithmetic dealt mainly with integral numbers, the properties of odds and evens and their compounds, prime numbers and their products, ratios and means.

It is not generally appreciated that later Greek mathematicians laid down the foundations of trigonometry and algebra. Trigonometry may be said to have started with a work, now lost, by Hipparchus of Nicaea and Rhodes (*c.* 190–after 126 BC) on the chords in a circle. Building on this, Ptolemy of Alexandria (flourished AD 127–48) proved in his *Mathematical Syntaxis*, better known as the *Almagest*, that in a semicircle with diameter 120, if E is the midpoint of the radius DC, and EF is made equal to EB, then FD is the side of a decagon and BF the side of a pentagon inscribed in the circle. From this he calculates (if

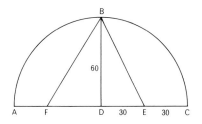

we may decimalise) that $EF^2 = BE^2 = 4500$, so $EF = 67.08204$. FD therefore is 37.08204, so its square is 1375.0777, while $BD^2 = 3600$, so $BF^2 = 4975.0777$ and $BF = 70.534$. So this gives us a value for FD/FB, which is the sine of angle FBD—that is, the sine of 36°. Other quantities may similarly be calculated.

Algebra was developed by Diophantus of Alexandria, who flourished *c*. AD 275. One equation emerges from the following problem. Find two numbers whose sum is 20 and the sum of whose squares is 208. Let their difference be $2x$; then the greater is $x + 10$, the smaller is $10 - x$. The sum of their squares is $2x^2 + 200$, which totals 208; hence $x^2 = 4$ and $x = 2$. So the numbers are 12 and 8.

We hear little of the school curriculum. Mathematics was a subject taught at every level, starting with simple counting and games. For counting, 1–9 were shown by three positions each of the last three fingers of the left hand, while the thumb and forefinger showed tens. The right hand showed hundreds and thousands. The system is explained by Bede in *De temporum ratione*, and by Nicolas Rhabdas of Smyrna, who flourished AD 1341. For educational play in the early stages of arithmetic Plato mentions the distributing of apples or garlands. For slightly older pupils money was introduced into the arithmetic, as at Rome (see chapter 8), together with other practical uses of mathematics, such as commercial, architectural, military and navigational applications.

In these connections the two notations mentioned above had to be taught thoroughly. Although the Greeks were great theorists, the development of town planning, for example, shows how the practical application of geometry was constantly before the eyes of pupils in many Greek cities as far apart as Asia Minor and Sicily. At the same time Pythagoras' mystical approach to numbers (1 = reason, 2 = opinion and the first female number, 3 = the first male number, 4 = justice, and so on) seems to have persisted. We may compare the Chinese equivalents in the *Ta Tai Li Chi* of AD 85–105: 1 = heaven, 2 = earth, 3 = mankind.

Examinations at Magnesia ad Maeandrum in the second century BC were in arithmetic, drawing, music, and lyric poetry. Music likewise featured in the Diogeneion at Athens, where geometry, literature and rhetoric were also taught. We can tell from other subjects that the question and answer method, which persisted long after classical antiquity, was common.

The equipment used in schools could consist of the abacus, the wax tablet, potsherds, used and unused papyrus sheets, books (also on papyrus until the third century AD), and instruments. The abacus probably reached Greece from Phoenicia, since its name is derived

13 Modern Chinese abacus, corresponding to one type of Graeco-Roman abacus. Each row, starting from the right, represents a power of ten. Beads above the bar are worth 5, beads below it 1. To show a number, beads are placed against the bar. The number shown here is therefore 6543.

from Old Semitic *abaq*, 'sand'. It originally consisted of grooves drawn in sand, in which pebbles were moved. The principle is that, as soon as a collection of pebbles can be replaced by one pebble elsewhere, that is done, either by taking them out of the grooves or, more simply, by pushing them to the 'dead' end. The Greeks normally wrote numbers, as we do, in descending order from left to right, so the right-hand groove (or one half of it) was for units and the left-hand groove was for the highest denomination. Whether the Greeks most often operated, as with Chinese abaci, in fives with a dividing line, or in fives without one, or in tens, is not known.

The wax tablet, single or multiple, was useful for written sums, which could easily be obliterated. There is evidence that some of these tablets had a hole so that they could be hung on the wall. The potsherd, *ostrakon*, was sometimes used for odd notes or calculations. For more permanent purposes sheets of papyrus, often imported from Egypt, were kept in stock. Some of those which have been found contain multiplication or conversion tables compiled by adults. A papyrus of the third century BC gives a list of squares, like the powers to be seen in a manuscript of a Latin geometry doubtfully ascribed to Boethius (fig. 14). Another papyrus, of the first century BC, has addition tables of monetary fractions. For simple additions the best specimens are on Coptic tablets of the late Roman Empire. Mathematical books may only gradually have been introduced into schools and only for the use of the teacher, not in multiple copies, which in manuscript took a long time to produce. School instruments included the ruler, compasses and perhaps double dividers. No protractors have been discovered, but certain types of set square could be brought into service for measuring right angles.

14 A manuscript table showing 1, 5 and 6 to the powers of 1 to 5 in Roman numerals. British Library, MS Harl. 3595, f. 26.

4 Measurement

All ancient civilisations used parts of the human body for many of their shorter measurements, while the longer ones reflect in origin the interests of their users. Thus fingers, palms, feet and forearms came to be standardised for the shorter units. The forearm was usually reckoned from wrist to elbow rather than from fingertip to elbow. As specimens of longer units we may mention, in ascending order, the double pace, the cord and the stade. The double pace originated in Rome, with 5 ft making 1 *passus*; the Romans thought of long distances in terms of land journeys on foot by their efficient road system, and of the marching speed of an army. The measuring cord was used by several early peoples, including the Sumerians and the Egyptians. The Greek *stadion* is in origin the length of a race-track, subsequently extended to a straight line, first on land and then over the sea. Since in the Graeco-Roman world the Greeks were the great sailors and the Romans land travellers, sea distances were usually measured in stades and land distances in miles of 1000 double paces.

Modern scholars have been able, in the case of all the major civilisations, to establish the average or customary measures of length. Such studies are complicated, in the first place, by the fact that there was in many areas a lack of uniformity—even in different parts of the Greek world. The same applies to units of volume and weight. Many sales, for example, must have been conducted by what we still call 'rule of thumb', though gradually official standards came to be laid down. In the Agora at Athens, weights and measures of the official inspectors (*metronomoi*) have been found. The task of working out equivalents is easiest where there are, as from Egypt, subdivided and inscribed rulers. Helpful too are the rare instances, from Greek temples in Asia Minor, where we have the equivalent of an architectural blueprint. More often we can deduce units from the dimensions of buildings, columns, etc., because round numbers of units were employed much more often than fractional ones. But there is always the problem that standards differed, sometimes slightly, sometimes greatly, according to region and period.

Egypt

Egyptian linear measurements were based on the royal cubit ('forearm'), of approximately 52.3 cm, whose subdivisions were the palm (the width of the palm excluding the thumb) and the digit (the finger's-breadth):

Measures of length

    4 digits (4 × 1.87 cm) = 1 palm, approx. 7.5 cm
    7 palms = 1 cubit, approx. 52.3 cm
    100 cubits = 1 *ḥt* or *khet* ('rod'), approx. 52 m 30 cm

Shorter and longer cubits encountered in Egypt appear to be non-Egyptian. Two other measures are the *nebiu*, 'wooden pole', somewhat larger than the cubit, and the *itrw*, 'river-measure', estimated as normally 200 *ḥt*, but less in one text.

15 Cubit measuring rod from Egypt, divided into digits (finger's-breadths), which are in turn subdivided into decreasing fractions from $\frac{1}{2}$ to $\frac{1}{16}$ of the digit. Turin, Museo Egizio.

Measures of area

1 cubit × 100 cubits is called 1 cubit, 27.35 m² (the Egyptians here ignored any word corresponding to 'square')
100 of these cubits = 1 square *khet* (*ḥt*), Greek *aroura*, literally 'arable land', in Egypt of the Ptolemies 2735 m²
Subdivisions of the square *khet*: $\frac{1}{8}$, $\frac{1}{4}$, and $\frac{1}{2}$ ($\frac{1}{2}$ square *khet* = *rmn*)
10 square *khet* = 'a thousand' (cubits as defined above)

Measures of capacity

32 *ro* (approx. 0.015 litre) = 1 *hin* (approx. 0.489 litre)
10 *hennu* (pl. of *hin*) = 1 *hekat* (approx. 4.89 litres)
20 or 16 *hekat* = 1 *khar*, 'sack' (approx. 97.8 or 78.2 litres)

The *hekat* was the official corn measure, and its subdivisions were in the series $\frac{1}{2}$ *hekat* = 160 *ro* to $\frac{1}{64}$ *hekat* = 5 *ro*, written in symbols known as the Horus eye. A legend had it that the god Horus had an eye torn apart in a fight, and these parts, later restored by the god Thoth, were depicted as in fig. 17. The connection between linear and capacity measure is shown in the Rhind mathematical papyrus, nos 41–47. Thus no. 41 runs: 'Example of working out a circular container of diameter 9 and height 10'. What is wanted is the amount of corn that will go into this cylindrical container. The corn is reckoned in *khar*, at the earlier equivalent of 20 *hekat* to the *khar*. The answer, here correctly given, is 960 *khar*, and the calculation shows that π is taken as $3\frac{13}{81}$, or approximately 3.1605. The volume is $\pi r^2 h = 3.1605 \times 4.5^2 \times 10 = 640$ cubic cubits; this is multiplied by $1\frac{1}{2}$ to give the answer, so 1 cubic cubit = $\frac{2}{3}$ *khar* with this equivalent.

16 Egyptian alabaster vessel inscribed with a stated capacity of $8\frac{1}{6}$ *hennu*, equivalent to about 4 litres. New Kingdom, *c*. 1300 BC. British Museum.

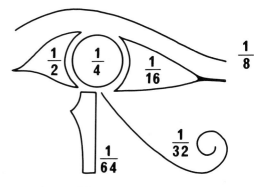

17 The eye of the Egyptian falcon god Horus served as a pictogram. To denote fractions of the *hekat*, used as a corn measure, scribes wrote in individual parts of the eye.

18 Division of the digit into 2 to 6 parts, from the statue of Gudea, king of Lagash, Mesopotamia. Paris, Louvre.

Mesopotamia

Sumerian measures of length

Mesopotamian linear measurements were based on the Sumerian cubit of 49.5 cm. This is known from a statue dating from 2170 BC of Gudea, king of Lagash (now in the Louvre, Paris). A plan of a building, incised on stone, lies on Gudea's lap. There is also on the statue a scale-bar with successive digits divided into 6, 5, 4, 3 and 2 equal parts. From this and other evidence we may give:

30 digits, of 1.65 cm each = 1 cubit (*kuš*), 49.5 cm

There is only very limited evidence, in Assyria in the second millennium BC, for use of the foot as a measure.

For surveying purposes the Sumerians used a unit called a reed, of 6 cubits, and one comparable with our pole, of 12 cubits. Representations of a coiled land-measuring cord and compasses are to be seen on the limestone stele (slab) of Ur-Nammu, king of Ur *c.* 2100 BC, now in the University of Pennsylvania Museum at Philadelphia. Since the accompanying text mentions canal-digging, we may conjecture that the cord and compasses were used for that purpose, though the building of the ziggurat of Ur required similar instruments.

Sumerian measures of area

1 square *gar* = 1 *sar*, approx. 36 m^2
100 (expressed as 1,40) *sar* = 1 *iku*
1800 (expressed as 30,0) *iku* = 1 *bùr*

The *sar* was used mainly for measuring the area of houses, the *iku* and the *bùr* for field areas; and there were other less common measures of area.

Sumerian measures of capacity

10 *sìla*, each of approx. 0.82 litre = 1 *bán*, approx. 8.2 litres
6 *bàn* = 1 *nigida*, approx. 49.2 litres
5 *nigida* = 1 *gur*, approx. 246 litres

A silver vase from Lagash, dating from 2450 BC, now in the Louvre, measures 4.15 litres, 5 *sìla*, up to the base of the neck.

A very early specimen of the Old Babylonian period, found at Tell al Rimah, is in the Iraq Museum, Baghdad. Its inscription is thus rendered: '1 homer 5 *sūtu* ⅓ *qû*, measured in the *sūtu* of Šamaš'. Since 1 homer = 10 *sūtu* and 1 *sūtu* = 10 *qû*, the jar held 150⅓ *qû*. Its measured capacity, within a margin of error of 2%, is 121.3 litres, so that 1 *qû* = 0.79–0.82 litre. The fact that quite different equivalents have been suggested for other vases is chiefly explained by the phrase 'measured in the *sūtu* of Šamaš', which clearly implies local variants. In the Neo-Babylonian period 1 *qû* is also the area of land that can be seeded with 1 *qû* or *qa* of seed.

19 Jar from Fort Shalmaneser, Nimrud, Mesopotamia, inscribed '1 homer, 3 *sūtu*, 7 *qû*'. British Museum, on loan from the University of London.

Greece
Greek city states varied appreciably in their measures.

Measures of length
The extremes were based on a foot (*pous*), varying according to region between 27 and 35 cm. Two measures used in temple construction were 32.6–32.8 cm and 29.4–29.6 cm, the latter being similar to the normal Roman foot (p. 27).

> 4 *daktyloi* (finger's-breadths) = 1 *palaste* (palm)
> 3 *palastai* = 1 *spithamē* (span between thumb and little finger)
> 4 *palastai* = 1 foot
> 1½ feet = 1 cubit
> 4 cubits = 1 *orguia*
> 10 *orguiai* = 1 *amma* (cord)
> 10 *ammata* = 1 *stadion*

Alternatively, 10 feet = 1 *akaina*, 10 *akainai* = 1 *plethron*, 6 *plethra* = 1 *stadion*.

Measures of area

> 10,000 square feet = 1 *plethron* (the *plethron* could also be a linear measure of 100 feet, as given above).

Measures of capacity
These also varied from place to place, but with the commonest equivalents were as follows.

For liquid measures:

> 6 *kyathoi*, each of 0.04 or 0.045 litre = 1 *kotylē*, 0.24 or 0.27 litre
> 2 *kotylai* = 1 *xestēs*, 0.48 or 0.54 litre
> 6 *xestai* = 1 *khous*, 'pourer', 2.88 or 3.24 litres
> 12 *khoes* = 1 *metrētēs*, 'measurer', 34.56 or 38.88 litres

The water capacity of a clepsydra (see chapter 7) found in Athens is 6.4 litres, clearly intended as the local value of 2 *khoes*.

For dry measures:

> 6 *kyathoi* = 1 *kotylē*, as above
> 4 *kotylai* = 1 *khoinix*, 0.96 or 1.08 litres
> 8 *khoinikes* = 1 *hekteus*, 7.68 or 8.64 litres
> 6 *hekteis* = 1 *medimnos*, 46.08 or 51.84 litres

Rome
The smallest Roman measurement, like the smallest Greek, was a finger's-breadth (*digitus*). As in Greece and elsewhere, 4 of these finger's-breadths formed a palm, and 4 palms made a foot, so that there were 16 *digiti* to a foot. But we also find the system which was to survive after the classical period. From at least the late second century BC the Latin duodecimal system was brought in to divide measures into 12 parts. The Latin word for one-twelfth, *uncia*, is derived from *unus*, so literally means 'unit'. It has given the English language both 'ounce' and 'inch' (Old English *ynce*).

The foot itself (*pes*) was normally of 29.57 cm, though in the provinces we also find a *pes Drusianus*, named after Nero Claudius Drusus (38–9 BC), stepson of Augustus, which had a length of 33.3 or 33.5 cm; it was really of far greater antiquity than Drusus. From the third century AD there was also a short foot of 29.42 cm. If one was dealing purely with linear measurement, then 5 *pedes* made 1 *passus* (literally pace, but in fact a double pace). One thousand *passus* (*mille passus*) made 1 Roman mile (1.4785 km). The abbreviation M.P. stands for this, or the plural *milia passuum*. Land surveyors used linear divisions of 120 *pedes* to make up 1 *actus* = 35 m 48.4. The *actus* (plural *actus*), from *ago*, 'drive', was in origin the distance that oxen pulling a plough would be driven before turning.

Measures of area

For areas of land, whether in agriculture or in land survey, the *iugerum* was the normal measure. The full table, based on the standard foot, is as follows.

14,400 square feet (*pedes quadrati*) = 1 square *actus* (*actus quadratus*), of about 0.126 ha
2 square *actus* (*actus quadrati*) = 1 *iugerum*, of about 0.252 ha
2 *iugera* = 1 *heredium*, heritable plot, of about 0.504 ha
100 *heredia* = 1 *centuria* of regular size (200 *iugera*), about 50.4 ha

The word *heredium* was little used in practice. The *centuria* varied in size from 50 to 400 *iugera*, though 200 *iugera*, as given above, with squares of size 20×20 *actus* = 709.68×709.68 m, was by far the commonest size.

Measures of capacity

For liquid measures:
4 *cochlearia*, 'spoonful' = 1 *cyathus*, approx. 0.0455 litre
3 *cyathi* = 1 *quartarius*, approx. 0.137 litre
2 *quartarii* = 1 *hemina*, approx. 0.273 litre
2 *heminae* = 1 *sextarius*, approx. 0.546 litre
6 *sextarii* = 1 *congius*, approx. 3.275 litres
8 *congii* = 1 *amphora*, approx. 26.2 litres, properly 1 Roman cubic foot
20 *amphorae* = 1 *culleus*, approx. 524 litres

For dry measures:
4 *cochlearia*, 'spoonful' = 1 *cyathus*, approx. 0.0455 litre
3 *cyathi* = 1 *quartarius*, approx. 0.137 litre
2 *quartarii* = 1 *hemina*, approx. 0.273 litre
2 *heminae* = 1 *sextarius*, approx. 0.546 litre
16 *sextarii* = 1 *modius*, approx. 8.736 litres

Thus Roman measures of liquid capacity were based on the equivalent 288 *cochlearia* = 1 *congius*. The *cochlear* was originally used for extracting snails from their shells. The *congius* was used chiefly for liquid measures, its place for dry measures being largely taken by the *sextarius*, so called because it was one-sixth of a *congius*. The commonest measure

20 Roman vase marked SEXTAR (*sextarius*). It contains not the usual liquid measure of 0.546 litre but 0.99 litre. British Museum.

21 Mosaic showing Roman slaves filling a
modius with corn brought from a ship. Ostia,
'Aula dei Mensores'.

22 Corn-measure from Magnis (Carvoran) on
Hadrian's Wall. The inscription reads: IMP ...
CAESARE AVG. GERMANICO. XV. COS. EXACTVS.
AD. Ƨ. XVII S HABET. P. XXXIIX ('under the
Emperor [Domitian] Caesar Augustus
Germanicus, consul for the 15th time, made
exact to $17\frac{1}{2}$ *sextarii*. Has weight 38 lb').
Northumberland, Chesters Museum.

for dry materials was a *modius*, usually translated 'bushel'; mosaics show slaves filling
cylindrical containers of that size with corn. The biggest measure for liquid materials was
the *amphora*, which took its name from the large earthenware vessel used to transport
wine or oil by sea. *Amphorae* in fact varied in size, but obviously a dealer could insist
on one of standard measure containing one cubic foot of wine or oil. Owing to mass produc-
tion, those found on any one ship tend to be of the same size, as can be seen from ships
wrecked at Albenga (province of Savona, north Italy), Antibes and elsewhere.

As an example of a Roman container the bronze corn-measure of Carvoran, Northumberland, may be cited (fig. 22). It dates from AD 90, and its inscription states that it contains
exactly $17\frac{1}{2}$ *sextarii* (symbol Ƨ) and weighs 38 *librae* (symbol P = *pondo*). This was dry
measure, but it was tested both with rape seed and with water, giving the result that 1
sextarius = 0.645 litre and 0.648 litre respectively. These figures are appreciably higher
than the 0.546 litre given above.

5 Mathematics for the Surveyor and Architect

The occupations to be considered in this chapter, together with their end-products, were concerned with roads, aqueducts, land division, military survey, town planning and architecture. We possess Latin manuals of land survey (*Corpus Agrimensorum*), camp-making (Hyginus) and architecture (Vitruvius). From these and other ancient works, together with archaeological remains, we are able to build up a picture of those activities, more complete in some spheres than in others.

Many Roman roads, and a few Greek, are well preserved. Typical features in the western provinces of the Roman Empire are their straightness, except where the geography suggested an alternative alignment, and the fact that changes in direction often occur at a rise and are made at an angle, not normally with a curve. Statius, in his poetic account in *Silvae* iv.3 of Domitian's road to Naples, makes it clear that the work-force was large and well organised. Since the *groma* was the instrument of Roman land surveyors, and since there was a '*groma* area' in military camps, many modern writers have supposed it was used for road survey too, but we cannot prove this. What we do know is that very long stretches of road were often made exactly straight, and that there is evidence of accurate planning. For example, Stane Street, which connected London to Chichester, has a straight stretch

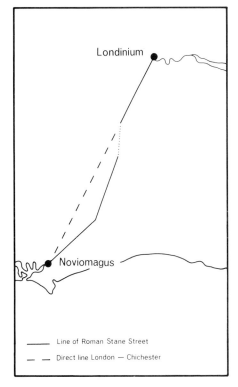

23 The course of Stane Street, the Roman road from London to Chichester.

24 Roman milestone from Rhiwiau-uchaf, south of Llanfairfechan, N. Wales, inscribed A.KANOVIO.M.P.VIII, '8 miles from Canovium (Caerhun)'. British Museum.

from near the Thames to the Epsom area. After that it changes direction, but the original line points exactly to Chichester. This kind of precision could have been achieved by the use of beacons, though some preliminary orientation could, as suggested by A. L. F. Rivet, have been carried out with the help of homing pigeons. The use of these and beacons for Graeco-Roman military intelligence is well attested. Pigeons based on London and Chichester could possibly have been released from the opposite ends of the route or from points on the way and their angles of flight recorded.

Main Roman roads in most areas had milestones (fig. 24), many of which record distances from nearby towns. Such distances are sometimes measured from the centre, sometimes from the outskirts. The unique Golden Milestone erected by Augustus in Rome recorded distances of places in Italy not from the milestone but from the Servian Wall. There is some evidence that road surveyors may have dealt with a one-mile section of road at a time. Research has been started on the angles taken by successive stretches of Roman road. This has suggested that divergences were more often than not in simple fractions of 180°, e.g.: 15° divergence = angle of 165°; 7½° divergence = angle of 172½°. Widths of Roman roads tended to be planned in round numbers of feet, e.g. 8, 12, 16, 20, 24, 30 or 40 feet.

Aqueduct surveyors needed mathematics particularly for tunnelling, and for calculating slopes and volumes of piped water. An early tunnel is that of Eupalinus at Samos (c. 530 BC), 1 km long. Hero of Alexandria (c. AD 60) shows how to start the digging of a tunnel from both sides of a hill with the correct orientation. His method is to measure lines at right angles round the side of the hill, then by means of similar triangles to establish the correct angles for starting the two sides (fig. 25). Evidently the advice was not always followed, since an inscription from Saldae, Mauretania (Bejaia, Algeria) shows that after local excavators had failed to meet in the middle, the Roman surveyor Nonius Datus was called in, made a map of the area, and planned correct lines of approach. For ensuring the proper gradient on aqueducts it was probably sufficient to align a series of poles of appropriate length, since Romans seem to have distrusted elaborate devices such as Hero's *dioptra* (fig. 26). The volume of water supplied depended on the diameter of the pipe or pipes, and detailed instructions for this are given by Frontinus in his work on the water supply of Rome. The positioning and use of siphons in aqueducts was a specialised skill.

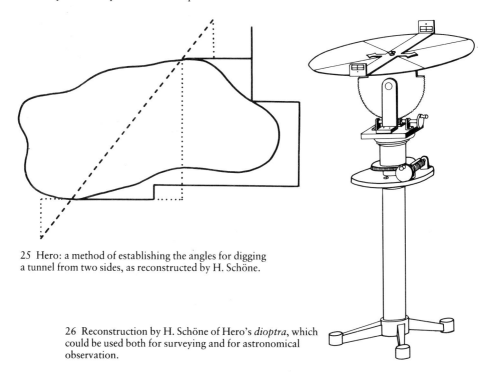

25 Hero: a method of establishing the angles for digging a tunnel from two sides, as reconstructed by H. Schöne.

26 Reconstruction by H. Schöne of Hero's *dioptra*, which could be used both for surveying and for astronomical observation.

We happen to know more about Roman land survey than other types because of the existence of the *Corpus Agrimensorum*, a collection of manuals of various dates, though traces of Roman land division in Italy, Tunisia and many other areas, together with inscriptions, have also made a substantial contribution. The standard 'century' was a square with sides of 20 *actus* or 2400 Roman feet. Owing to local variations the actual measurement comes to between 703 and 710 m. Since 2 square *actus* make 1 *iugerum*, this gave the area of the standard 'century' as 200 *iugera*. Various other measurements, such as 50, 210, 240 or 400 *iugera*, are occasionally found. The land was divided by *limites*, literally meaning 'balks' between fields, but in practice these were either roads between 'centuries' or subsidiary tracks. Those in one direction were called *kardines*, those at right angles *decumani*; the two intersecting main roads were called *kardo maximus* (K.M.) and *decumanus maximus* (D.M.). The system of reference used to number 'centuries' started thus:

S.D. *sinistra decumani*	to left of *decumanus*
D.D. *dextra decumani*	to right of *decumanus*
C.K. *citra kardinem*	to near side of *kardo*
V.K. *ultra kardinem*	beyond *kardo*

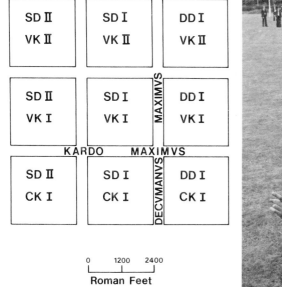

27 Method of numbering 'centuries'.

28 Roman surveying exercise by students, using a *groma*, small poles and *decempedae*.

The main instruments used for dividing up the land were the *groma* (cross-staff) and the *decempeda* (10-foot rod). The *groma*, as can be seen from the Pompeii model in the Naples National Museum, consisted of horizontal cross-pieces mounted on a bracket swivelling on a pole, each arm of the cross having a plumb-line and plumb-bob. Its main use was to survey squares and rectangles, with a back-check every five *limites*. The *actus* of 120 feet was measured, we may take it, with two *decempedae* end-on, care being taken to keep the line straight. When we tried measuring 2 *iugera* with university and school students, using replicas of the *groma* and *decempeda*, the resulting error was minimal.

Most Roman land division was carried out by *agrimensores*, literally 'land measurers'. Under the Roman Empire these came to be organised under a bureaucratic system, apprentices being taught about use of instruments, measurement and allocation of land, boundaries, mapping, land law, and the elements of mathematics and cosmology. The land which they divided was normally *ager publicus*, state land, attached to a colony. Roman colonies were old or new settlements with adjacent territory, designed either for veterans (ex-servicemen) or for civilians, either in Italy or elsewhere in the empire. There was inflation in the area of holdings allotted, starting from 2 *iugera* and common pasture, and ending with $33\frac{1}{3}$–100 *iugera*, veterans being given plots according to rank. If necessary, a holding could be distributed between two or three 'centuries'. When the allocation had been made by lot, the surveyor took settlers to their holdings and made a map of the land division.

Surveyors, like farmers in Columella's *Agriculture*, were taught how to calculate areas. Some calculations were straightforward and exact; some were approximations depending, for example, on averaging out two sides. Columella's figures relate more to theoretical geometry than to likely field shapes. In the exact category we find a square, a rectangle, a trapezium, and a right-angled triangle. The area of the square is expressed as $\frac{1}{3}$ *iugerum* (9600 ft^2) + $\frac{1}{6}$ *uncia* (400 ft^2). The rectangle illustrates standard land division procedure. The trapezoidal field comes into the exact category because Columella evidently had in mind a regular trapezium. In the approximate category, for calculating the area of an equilateral triangle Columella uses $\sqrt{3} = 1.7\dot{3}$ and for the area of a circle $\pi = 3\frac{1}{8}$. For a segment he uses the same value for π, but his instructions work only where the angle subtended at the centre of the circle is a right angle.

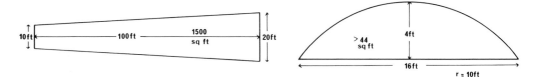

29 Reconstruction of Columella's calculation of the area of a regular trapezium and a segment of a circle.

Triangles were not used for regular survey work, as they have been since Tudor times. But the right-angled triangle and its properties formed part of the basic training, though problems were sometimes unpractical. One given in the *Corpus Agrimensorum* assumes that in a triangle ABC with a right angle at B, we know the area and hypotenuse and the total AB + BC; we have to establish AB, BC separately. It is solved thus:

AC^2 = 289; subtract 4 × area (240), leaving a remainder of 49; take the square root, 7; add the sum of the two sides, 23, giving a total of 30; halve this, giving 15 = BC.

This can be shown to give the correct result.

M. Iunius Nipsus' use of triangles for measuring the width of a river was shown in chapter 2. Another use of triangles was in re-survey, where a faulty line had to be redrawn; but unfortunately the manuscript diagrams are corrupt.

Military surveyors had to be equally precise. We know of a line 29 km long on the German *limes* (in this sense 'boundary of the empire') whose variation from the straight is no more than 2 m. Hyginus, in his work on Roman camp-making, gives exact measurements of all the buildings, as well as outer defences, of a camp intended as a pattern. Military surveyors had also to measure sections of wall built. This was done in the case of Hadrian's Wall, the only complication being that some army units measured in miles and *passus* (double paces), while others measured in miles and feet. The abbreviation for miles, M.P., is unambiguous, since it was used only to mean *milia passuum*. But P. alone was ambiguous, since it could refer to *passus* or *pedes*.

Town-planning in the classical world started with Hippodamus of Miletus. After the end of the Persian Wars in 479 BC he made designs for his own city and a number of others, laying them out in rectangular blocks. In 443 BC he was one of the original settlers at the Athenian colony of Thurii, near Sybaris (Sibari) in southern Italy. As a result, we find that many Greek settlements in southern Italy and Sicily conform to this type, consisting of rather long rectangles. The present town centre of Naples, with its constant traffic jams, must be built on the same lines as the ancient Greek colony. To what extent the surrounding countryside of these colonies was divided into rectangles is not quite so certain.

One important task of the town surveyor was to make plans of the area. These were clearly less common than they are today, but we are fortunate in possessing numerous fragments of the *Forma Urbis Romae* (fig. 30), a marble plan of the city of Rome made between AD 203 and 208. A reconstruction of all fragments known in 1959 is to be seen in a courtyard of the Capitoline Museum, Rome. Geometrically, the plan clearly presented a challenge, because most of Rome had not been planned but had grown up in a haphazard manner. No measurements are given, but an attempt is made to work to a scale of 1:240 or 250. In actual fact prominent temples tend to be shown somewhat larger than they really are. Nevertheless it can be claimed that as a plan of Rome to scale this was not surpassed until the eighteenth century.

Architecture was one of the most highly esteemed arts in ancient Greek and Roman cities, and it is clear from Vitruvius' manual that the good architect had to be proficient in mathematics and measurement. It seems to have been more common for architects to make models than drawings of intended buildings, in this way introducing a three-dimensional effect. Much effort was spent on achieving perfect proportion in temples and other buildings; Vitruvius even advocates desirable proportions for the reception rooms in a house. A notable feature of the Parthenon on the Athens acropolis is what is known as 'upward curvature'. As D. S. Robertson has written: 'The stylobate [top step, literally "support for columns"] often drops towards all four angles, so that its surface is somewhat like that of a carpet nailed at its corners only, and raised from the floor by a draught.' If, on approaching the temple from either end, one places a hat on one corner of the stylobate and looks towards it with one's eyes near ground level at the other corner, one will not

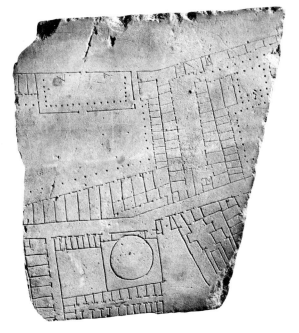

30 Section of the *Forma Urbis Romae*, a town plan of Rome, showing part of the Trastevere, with the Tiber at the top. Rome, Musei Capitolini (original in the Museo di Roma).

be able to see it. The architect has built in a slight convex curve, just as, by what is known as entasis, the columns often do not have a straight surface, but diminish in diameter by a very gentle convex curve, not so great as to cause a bulge. Whether, as the ancients maintained, the purpose was to correct the optical illusion that angle columns looked too thin, or whether there were more practical purposes, such as drainage in the case of the stylobate, is disputed. In the case of the columns, stone frequently replaced earlier wood, and an attempt may have been made to imitate the shape of the timber.

A building known to have been specifically designed by an architect called Polyclitus (younger than the famous artist of that name) was the theatre at the healing shrine of Epidaurus, whose auditorium is now thought to date from about 300 BC (fig. 31). The main object there was to ensure perfect acoustics; and although most of the stage buildings have long since disappeared, the auditorium is so well preserved that the acoustics remain extremely good. From the point of view of numerology, is it only a coincidence that the numbers of rows of seats below and above the *diazoma* (passage across the auditorium) are respectively 34 and 21? These are two successive numbers in the Fibonacci series. The series begins 1, 2, 3, 5, 8, each term being the sum of the previous two numbers. The ratio of successive numbers is a constantly closer approximation to the Golden Number

$$\frac{\sqrt{5}+1}{2}$$

(Thus $34/21 = 1.6190\ldots$, compared with the true value of the Golden Number, $1.6180\ldots$.) Where, as at Megalopolis in Arcadia, architecture could in a new town be combined with town-planning, full use was made of the space available.

Architects used either models, as mentioned, or scale plans to act as blueprints. A scale plan of the uncompleted temple of Apollo at Didyma, south of Miletus in Asia Minor, has long been visible in the ruins of the temple, but has only recently been interpreted. It consists of lines up to 10 m long, circles, etc., cut into the stonework in a passage. From the point of view of measurement, the most interesting feature is a blueprint designed to show the entasis on columns. Half the width only is shown, but at full scale. The height, on the other hand, is at $\frac{1}{16}$ size, so that each Greek foot is represented by a *daktylos* (digit or finger's-breadth). Similar but less important incised figures have been found at two other temples in Asia Minor, and it is to be hoped that more will be discovered. The concept of scale is important when put side by side with the development of mapping, to be considered in the next chapter.

31 The Greek theatre at Epidaurus.

6 Mapping and the Concept of Scale

Attempts to measure the earth and map it had been made earlier in many countries, but it was the Greeks of the classical and Hellenistic period who pioneered advances in this field. A distinction was made between the earth and the inhabited earth (*oikoumene*). The continents known to the Greeks and Romans were Europe, Asia and Africa, also called Libya. To Hecataeus of Miletus (*c*. 500 BC) Africa was part of Asia, though others separated them (fig. 32). None of the three continents was fully known, but it was generally agreed that the east–west length of the inhabited world was about $1\frac{1}{2}$ or 2 times its north–south width. Attempts were then made to measure a north–south and an east–west line, using as unit the stade (*stadion*), 1 stade being between 150 and 200 m. Dicaearchus of Messana (Messina), a pupil of Aristotle writing about 320 BC, divided the known world by a west–east line from the Straits of Gibraltar via the Straits of Messina (7000 stades) and the Peloponnese (a further 3000) to southern Turkey and the Himalayas.

Eratosthenes of Cyrene (*c*. 275–194 BC), director of the Alexandrian library, measured dividing lines of the inhabited world intersecting at Rhodes, a centre of Greek navigation (fig. 33), and approximately 74,000 stades in length east–west, 38,000 north–south (what the ancients called 'breadth'). Measurement of the whole earth became possible once it was assumed, as is not far wrong, that it is a perfect sphere. After such assumptions by Pythagoras (flourished 530 BC) and Parmenides (born *c*. 515 BC), Aristotle (384–322 BC) was able to suggest a circumference of 400,000 stades (60,000–80,000 km).

A far more accurate measurement was made by Eratosthenes. He observed that at Syene (Aswan, Upper Egypt) the sun at mid-day on the summer solstice was exactly overhead, whereas at Alexandria it formed an angle with the vertical of $\frac{1}{50}$ of 360°. He then assumed that: (a) the sun was so distant that one could consider its rays to be parallel anywhere on earth; (b) Syene was on the same longitude as Alexandria (there is actually a difference of 2°); (c) Syene was 5000 stades from Alexandria. This resulted in a circumference of $50 \times 5000 = 250,000$ stades, which he adjusted to 252,000 in order to make it divisible by 60.

32 Conjectural reconstruction of Hecataeus' map of the known world.

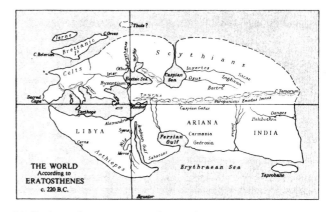

33 Conjectural reconstruction of Eratosthenes' map of the known world.

If Eratosthenes was then using a stade of 178.6 m, his amended circumference comes to about 45,000 km, as against the actual equatorial circumference of 40,075 km. But one theory has it that he was using an Egyptian land measure roughly equated with a stade and having a value of 157.5 m, in which case his estimate was much closer, at 39,690 km.

Knowledge of the geometry of the sphere enabled Greek mathematicians to reckon the approximate number of stades separating any two places on a known parallel. Thus, if the circumference at the parallel of Rhodes, 36° N, was reckoned as 195,000 stades (Eratosthenes wrongly considered it less than 200,000), one degree of longitude would there be equal to $541\frac{2}{3}$ stades, as against 700 stades on the equator.

The circumference suggested by Eratosthenes was criticised first by Hipparchus (flourished 162–126 BC) as too small, then by Posidonius (115–51/50 BC) as too large. Posidonius established that the star Canopus was on the horizon at Rhodes at the same time as it reached an elevation of $\frac{1}{48}$ of 360° at Alexandria. If the two were on the same longitude (they are actually 1° 37′ apart), he had only to multiply their distance from each other by 48. But this sea distance was harder to calculate than one on land; and after first making it 5000 stades, he settled for 3750, which gave a circumference of 180,000 stades, each degree of longitude at the equator being 500 stades. Such a round sum appealed to Ptolemy (c. AD 90–c. 168), who claimed that the inhabited world extended over 180° of longitude, not, as Marinus had given, 225°. At the latitude of Rhodes he made this 180 × 400 = 72,000 stades, from the Canaries to the coast of China. Ptolemy's figure of 400 stades for each degree of longitude at the latitude of Rhodes is a slightly different proportion from that of Eratosthenes.

The projection of world maps ran into difficulties as more came to be known about the inhabited world. By Ptolemy's time this was recognised as extending south of the equator. He tried to provide for a map of the inhabited world extending from 63° N, the latitude of Thule island (the Shetlands), to 16° 25′ S, Agisymba and Prasum promontory in East Africa. Ptolemy suggested two different projections (fig. 34). The first, a conic projection, had straight meridians diverging from 63° N to the equator, after which they turned abruptly and contracted southwards to 16° 25′ S. The second had curved meridians designed to have 'the shape which they appear to have on a globe'. For this purpose the latitude of Syene (Aswan), 23° 40′ N, about half-way from north to south of his inhabited world, was taken as the central parallel. This is a difficult projection to draw, since the nearer the parallels are to the sides of the map, the more curved they have to be made. Also it was less easy to locate places with known latitude and longitude on such a world map. So for practical purposes the first projection was recommended. (A third which he outlined was probably not intended to be seriously considered by map-makers.)

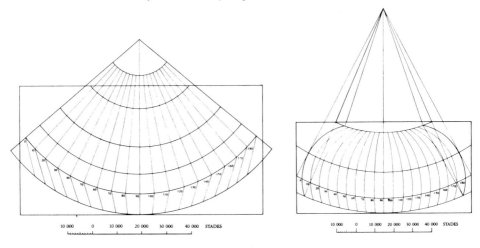

34 Ptolemy's first and second projections for a world map.

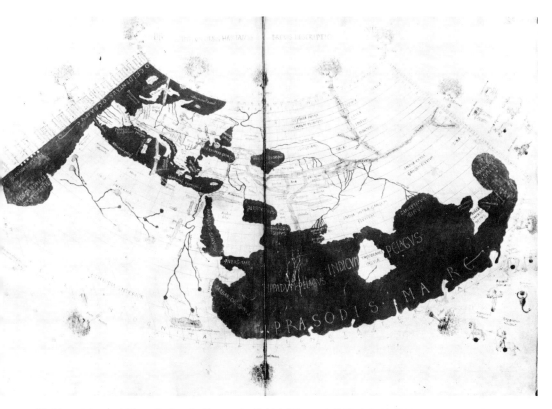

35 Map of the world from Ptolemy's *Geography*. British Library, MS Harl. 7182, f.58v–59r.

For regional maps, on the other hand, Ptolemy proposed straight lines of latitude and longitude intersecting at right angles. This made it easier to insert place-names from given lists of latitude and longitude. But in order to provide the approximately correct proportions, each regional map was given its own representative fraction, based on its mean latitude. Ideally these proportions should result in distortion being reduced to a minimum; but in fact Ptolemy, as can be seen from his map of northern Britain, not infrequently made mistakes which resulted in imperfections despite the correct regional proportions.

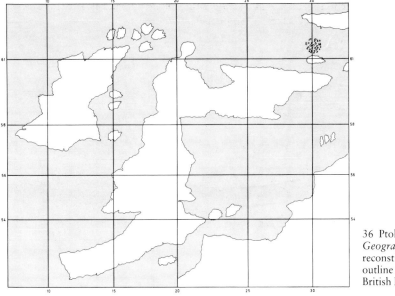

36 Ptolemy's *Geography*: reconstructed outline map of the British Isles.

Even in Ptolemy's *Geography* we find two methods of calculating latitude and longitude. The standard method of giving latitude is our own method, but in addition we find another which gives for each place in the better-known areas of the world its maximum hours of daylight, rounded off to the nearest quarter-hour. The standard method of giving longitude is by degrees east of the Canaries. But we also encounter a method of giving the distance east or west of Alexandria in hours (since $360° = 24\,h$, $15° = 1\,h$).

Whereas the ancients were able to calculate latitudes fairly well, longitudes presented greater problems. It was known that these could be calculated by simultaneous observation of an eclipse from two or more places. The problem was that scientific data were lacking for such an attempt. An astronomical method used by Hero of Alexandria to calculate the distance between Rome and Alexandria, although based on scientific principles, did not result in an accurate answer. Under the late Roman Empire, Greek mathematicians, using trigonometrical methods, seem to have given distance equivalents for some places based on their latitudes and longitudes in Ptolemy's *Geography*.

In addition to terrestrial mapping, the positions of visible stars were carefully plotted with similar co-ordinates. By this method Hipparchus listed the co-ordinates of 800 stars. Ptolemy in his *Almagest* gave the co-ordinates of 1022 stars whose positions could be plotted on a globe. Archimedes designed an instrument like an orrery, which portrayed the motions of the sun, the moon and five planets, even showing solar and lunar eclipses after the correct number of revolutions.

The aspect of measurement is confined to the written word in the Peutinger Table (fig. 37), sole surviving copy (now in the Austrian National Library, Vienna) of a road map of the Roman Empire. It was earlier in the form of a single parchment roll, 6.75 m long but only 34 cm wide; but it was cut up and placed under glass for better preservation. It has north at the top; and north–south distances, such as that between the Italian and north African coasts, look very short, while east–west ones look very long. Many road distances, in Roman miles or in local measurements such as leagues, are written in, but there is hardly any attempt at scale. That a road map more or less to scale could exist in the Graeco-Roman world is shown by the Dura Europos shield (fig. 38), an unofficial map of the north shore of the Black Sea, with place-names written in Greek but with distances expressed in Roman miles; it is now in the Bibliothèque Nationale, Paris.

In addition to road maps, we possess from classical antiquity a number of itineraries and *periploi*. Itineraries gave details of road journeys in list form and were evidently commoner than road maps. The most helpful one for the reconstruction of ancient topography

37 Section of the Peutinger Table, a copy of a Roman road map, showing Dalmatia, Italy from Bolsena to Rome, and part of Tunisia (Carthage is opposite Rome's harbour at Ostia). Vienna, Österreichische Nationalbibliothek. (See also front cover.)

is the Antonine Itinerary, drawn up for journeys to be undertaken by an emperor of the Antonine dynasty. The most probable emperor is Caracalla, who journeyed via Asia Minor to Egypt in AD 214–15. Britain has 15 *itinera*, some going out of their way to enable the emperor to visit particular places. Iter III, from London to Dover, is expressed as follows:

London to the port of Dover, 66 miles, checked:
To Rochester [Durobrivae]	27 miles
From there to Canterbury [Durovernum]	25 miles
From there to Dover harbour	14 miles

Periploi are sea itineraries, sometimes containing supplementary information. They tended to be expressed in stades rather than miles. It is often unclear whether the distance recorded is the straight-line distance from cape to cape or that following the coast.

Most of the large-scale plans from the Roman world were drawn up by architects or surveyors, and as such have been mentioned in the previous chapter. But there are also other items, such as a mosaic plan of some baths, found at Rome, which has measurements of rooms in Roman feet; and the 'Urbino plan' (fig. 39; now in the Ducal Palace, Urbino), of a property recorded as measuring $546 \times 524\frac{1}{2}$ Roman feet, with an approach road 1688 Roman feet long. Many such items must have been destroyed as no longer relevant.

It is evident that throughout the classical period the concept of scale was well developed in professional work. Plans to scale can be seen from Babylonian times onwards. The special attention paid by Egyptian mathematicians to pyramids must have been supplemented by working models. Architectural plans at Didyma show a skilful blend of full size and $\frac{1}{16}$ scale; and the use, attested by Greek writers, of architectural models implies three-dimensional accuracy. Greek world maps from the third century BC onwards were based to a greater or lesser extent on spherical projection theory, which in its application to cartography reached its high point with Ptolemy. The Roman land and building surveyors clearly had a firm grasp of scale. The only difficulty lay in the variety of units between different areas or periods. After centuries when the city-states were independent, domination first by Alexander the Great and his successors and then by Rome led to a much greater degree of standardisation.

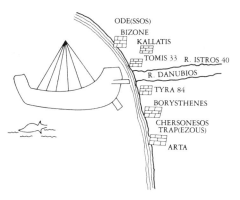

38 Simplified redrawing of the Dura Europos shield map. The numerals indicate distances, in Roman miles, from previous places westwards.

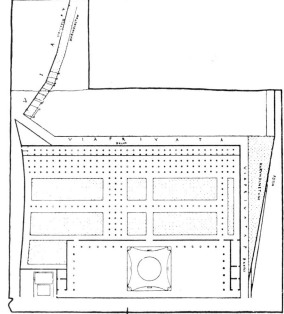

39 The 'Urbino plan' of a Roman estate.

7   Telling the Time

Near Eastern Calendars

The customary method of denoting years in the ancient Near East was by reference to the number of years that the current monarch had reigned. The Babylonians had a year which started in the spring and contained either 12 or 13 months, depending on the relation of the lunar months and the solar year. Their system and nomenclature were taken over by the Hebrews after 587 BC. By customary observance in the Babylonian world, a new month started, if the new moon was visible, on what would otherwise be the 30th day of the current month; if not, it began on the following day. Hence months were either 29 or 30 days long. An extra (intercalary) month was inserted as and when required, to bring the months into step with the sun.

Official Egyptian months had 30 days each, and the daily income of a temple was reckoned as $\frac{1}{360}$ of its annual revenue; but in fact 5 supplementary days were added to make 365. But there was still a discrepancy from the true length of the year, which is more nearly $365\frac{1}{4}$ days. So the new year, which fell on the 1st of the month Thoth, gradually fell back: in 747 BC it was on 26 February, in 647 BC on 1 February, in 547 BC on 7 January, and so on. The scribes were presumably trained to take account of this discrepancy.

Greek Calendars

Each city-state in classical Greece had its own method of expressing years, months and days. In many cases we can be confident, or fairly so, which year is intended, and can relate it to our era—e.g., 480 BC for the battle of Salamis. But the Athenians were governed by an annually changing *archon eponymos*, office-holder of the year, so that they tended to refer to a year as 'in the archonship of Euclides', etc. This system was not easily understood by other Greeks, let alone modern readers. Hence other systems evolved for referring to years:

(1) The Olympic games, the great occasion every four years when all Greeks met, were used from the fourth century BC as a system of reference. Since the games started in 776 BC, that year counted as Ol. (Olympiad) 1.1, Ol. 1.2 being 775; Ol. 2.1 (the year of the second games) was 772 BC, and so on. The system lasted down to the Byzantine Empire.

(2) A method recommended by Eratosthenes was to start at the traditional date of the fall of Troy, 1183 BC. He expressed it as 860 years before the death of Alexander the Great (who died in 323 BC).

(3) Eusebius (*c.* AD 269–340), the ecclesiastical historian who was born in Palestine and became bishop of Caesarea but who wrote in Greek, went back still further. He started from the birth of Abraham, which he reckoned as 2016 BC. By his time the Graeco-Roman world had become interested in the events of the Old Testament, and Eusebius was keen to refute the charge that Christians followed Judaism merely to change it.

Athenian months, which began at the summer solstice, were named after festivals, and lasted 30 or 29 days alternately. To bring the number up to an average of $365\frac{1}{4}$ days, an extra month had to be inserted when required, though this was not done very consistently. There is evidence that Meton's system (see p. 18) was not adopted. Other Greek cities had similar methods, but less is known of most of these.

Days of the month varied between Greek cities so much that Aristotle's pupil Aristoxenus says musical scales are as disputed as Greek calendars: 'The tenth day of the month at Corinth is the fifth at Athens and the eighth somewhere else.' Days of the month at Athens were reckoned partly from its beginning, partly from one-third of the way through it, and partly from its end. Thus day 1 was called 'new month'; days 2–10 were the 2nd–10th of the 'rising' month; days 11–19 were the 1st–9th after the 10th; the 20th was *eikas* (*eikosi* = 20); while days 21–29 were more often counted from the end of the 'waning' month than as '1st after *eikas*', etc. The 30th of the month was often called 'old and new [day]'. But from these phrases, originally based on the phases of the moon, it must not be supposed that an official month and a lunar month necessarily coincided.

In addition, the Athenians had a particular system of dating for meetings of the assembly. The number of days during which a particular tribe presided over the council was called a prytany. So a calendar of prytanies was adopted for this purpose, the formula being 'on day 1 (etc.) of the first (etc.) prytany', with the name of the presiding tribe.

The Roman Calendar

The Romans' normal method of dating in the classical period was to name each year after the two consuls of that year. These could be followed, if required, by a numeral: e.g., III = *tertio* or *tertium*, 'consul for the third time'.

A more scientific method was put forward at the end of the Republic. This depended on the traditional date of the founding of Rome, reckoned back from known dates: in our usage it works out as between 754 and 751 BC but most often as 753. The phrase used for 'from the foundation of the city' was *ab urbe condita*, abbreviated to A.U.C. This chronology was still used, out of antiquarian interest, in some Renaissance maps, seen for example in the Galleries of the Vatican. It is also found, with or without the BC / AD equivalent, in some text-books up to about 1900.

Under the Roman Empire, an emperor could be appointed tribune of the people for a year or for successive years. This, conferring a special power, had a prestige value because it pointed to continuity from powers granted under the Republic. Dates could be reckoned from this 'tribunician power'. Thus, especially in inscriptions referring to an emperor, we find phrases like TRIB(unicia) POT(estate) VIII.

Finally, about AD 525–540 the Christian monk Dionysius Exiguus introduced the method of dating by years of the Christian Era. This took its year 1 as the year of Christ's birth. But the actual date of that event was a few years earlier than Dionysius calculated. Since he started from year 1 and not year 0, the previous year is now called 1 BC; so if, for example, we want to calculate the bimillenary of some event which occurred in 10 BC, it should be in 1991, not in 1990.

Roman month names have lasted up to the present, but are not self-explanatory. Those from September to December obviously mean '7th–10th month', whereas they refer to the 9th–12th months. (There was even an unscrupulous tax-collector in Gaul in the first century BC who tried because of this to extort two months' extra taxation.) The reason is that March was originally the first Roman month. In fact, we are told that under the kings (traditionally 753–510 BC) there were only ten named months, March–December. The names January and February (from Janus, god of gates, and *februa*, expiatory offerings to the gods) were introduced at the beginning of the Republic, traditionally 509 BC; and in 153 BC January was made the first month. March to June were named after deities and religious festivals. July and August were introduced in the post-Republican period in honour of Julius Caesar and Augustus; previously, as Quintilis and Sextilis, these months had been related to their original position in the year, as fifth and sixth months respectively.

March, May, July and October had 31 days, February had 28, other months 29. To bring the total up from 355 days to $366\frac{1}{4}$ (one day too many, but evidently regarded as correct in early times), an intercalary month of 22 or 23 days was inserted after 23 February, at the end of the festival Terminalia, when decreed by the Pontifex Maximus. Macrobius comments on the reason why (before 153 BC) it was not inserted at the end of the last month, February: 'I suppose this was due to some religious taboo, so that March should not fail to follow February.' This was an unsatisfactory state of affairs, especially when the chief priest did not decree an intercalation that was due. Accordingly in 46 BC Julius Caesar proclaimed a 'once for all' year of 445 days to make amends, and gave the months their present lengths. He ruled that in leap years a day should be inserted after 23 February.

The Romans reckoned inclusively, that is, counting in both the day itself and the festival before which it was calculated, so that the 13th was described as 3 days, not 2, before the 15th. Roman days of the month were numbered inclusively before the Kalends, Nones and Ides. The Kalends were the first day of the month; the Nones and Ides were originally market days, the former being the ninth day (*nona*), by inclusive reckoning, before the Ides. In the original four long months, March, May, July and October, the Nones were defined as the 7th of the month; in the remaining eight months they were the 5th. The Ides, *idus*, were defined as the 15th of the original long months (hence Ides of March = 15 March) and the 13th of all other months. Some specimen dates, with abbreviations, are:

Abbreviation	Expanded form	Date
KAL. IAN.	kalendis Ianuariis	1 January
PRID. KAL. IAN.	pridie kalendas Ianuarias	31 December (day before Kalends)
A.D. IV NON. IAN.	ante diem quartum nonas Ianuarias	2 January (4th day inclusive before Nones)
A.D. III ID. MART.	ante diem tertium idus Martias	13 March (3rd day inclusive before Ides)

The Hours

In the Near East and in the world of Greece and Rome, two systems of reckoning hours were recognised. The first, known to dynastic Egypt, consisted, like our modern one, of 24 equal hours. These were called in Greek *hōrai isēmerinai*, 'equal-day hours', and were used in antiquity for scientific purposes, especially for astronomical observation. For everyday purposes, on the other hand, the ancients preferred a system of *mere* or *hōrai kairikai*, 'seasonal hours', which Herodotus attributes to the Babylonians. In this system the period of daylight was divided into 12 hours, so that the length of such an hour was greatest in midsummer and smallest in midwinter. At the latitude of Rome the longest and shortest seasonal hours were about $1\frac{1}{4}$ hours and $\frac{3}{4}$ hour respectively. The Egyptians divided the night into 12 hours, but the Romans divided it into watches. Naturally, such a system did not recommend itself to exactness, though, as will be seen, it was reasonably well suited to current methods of timekeeping.

Sundials

It is possible to tell the time at night from the stars, and this may have been done in the Near East, but no instruments for this purpose have survived. The sundial, on the other hand, dates from remote antiquity, and many have survived, though often not intact. The earliest from Egypt is a shadow-stick of the eighth century BC. It has a straight base with a raised gnomon at one end, which was set to face east in the morning and west in the afternoon; the base was marked with six divisions for hours. A sundial devised by Anaximander and set up at Sparta was presumably more sophisticated. The Greeks clearly became more knowledgeable than the Romans about the workings of sundials. One captured by the Romans at Catania, Sicily, in 263 BC was set up in Rome, where for 99 years either no one realised, or no one could persuade the authorities, that it was incorrectly set for the latitude of Rome.

The largest sundial of Graeco-Roman antiquity, which was of the shadow-stick type, was set up by Augustus near the Pantheon in Rome. Its gnomon was an Egyptian obelisk, still in the same area today, though moved from its original site. Egyptian obelisks were usually dedicated to the sun-god, and at least some of them may have been used in sundials in Egypt. Parts of the large Greek inscription of Augustus' sundial have recently been excavated.

An analysis of Greek and Roman sundials by Sharon L. Gibbs has shown that the largest number of those preserved are vertical ones of the type known as conical. Their concave face is part of the surface of a cone; they were designed to face south and had a vertical gnomon. Another well-represented vertical concave group is known as spherical, most being in fact hemispherical. A third group, which may be vertical or horizontal, is called planar, since it has a flat dial. The horizontal type has eleven hour lines, the equinox and the summer and winter solstitial curves. There is also a small group of vertical cylindrical concave sundials. One such of unusual shape was found in 1975 as far away as Ai Khanoum, Afghanistan; it seems to be set for a place on the tropic of Cancer, so could have been picked up as a souvenir, perhaps at Syene (Aswan).

For travelling, it was possible, from about the third century AD, to use a portable sundial. Several specimens of these have been found, though in some cases subsequently lost. One,

40 Conical sundial with hours marked in Greek letters, found in Alexandria. British Museum. Compare fig. 53.

44

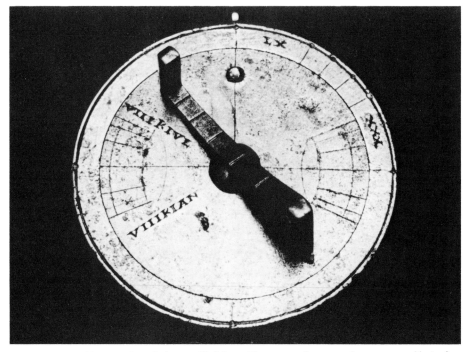

41 Roman portable vertical sundial: assembled front. The reverse shows province names and latitudes. (To reveal the inscriptions, the button is unrealistically set to the north pole and the gnomon outside the date range.) Oxford, Museum of the History of Science.

said to have been found in Bratislava but now in the Oxford Museum of the History of Science, is in bronze and has its inscriptions in Latin. It consists of a gnomon and an inner and outer dial. On the back are province names and mean latitudes. For Britain one finds the latitude 57°, too far north, as in Ptolemy. To set it in Britain one moves the inner dial so that the button is aligned with the 57° point: only XXX and LX are marked, but lines show 40° and 50°. One sets the gnomon to the appropriate date between the solstices, then suspends the sundial and turns it until the shadow from the gnomon falls squarely on the curved hour recorder. Automatically the gnomon adjusts to the height of the sun in the sky and points south at mid-day. The line on the recorder nearest to the gnomon indicates Roman hours 1 and 11, the next line 2 and 10, and so on.

Clocks
The sand-timer was no doubt the earliest timekeeper independent of the celestial bodies, but it is inefficient for measuring more than a limited duration. Hence attention turned at an early date to a simple form of water-clock. In dynastic Egypt a stone bowl with a small opening at the bottom was used, and specimens of these have been found. Since for everyday purposes the Egyptians used unequal hours, seasonal adjustments were made by reading the level against an appropriate set of lines on the side of the bowl. Similar simple clocks were used in other parts of the Near East.

Remains have been found of two types of Greek clepsydra, literally 'water-stealer'. One consisted of an upper cistern and a lower one into which water trickled. The level of water in the upper cistern was kept constant to avoid any change of pressure. The lower cistern had graduations on the sides to show the time. Ctesibius of Alexandria (flourished 279 BC)

42 Fragment of an Egyptian water-clock: interior. The dots indicate the level of the water at hourly intervals, and the figures at the bottom are amuletic symbols. The name of Alexander the Great is inscribed in hieroglyphs on the other side. British Museum.

devised the first clepsydra that could be called accurate. According to Vitruvius, it involved the use of a tube of gold to avoid clogging, an elaborate system of geared wheels, a cork drum, a type of siphon, moving figures, revolving pillars, falling pebbles or eggs, sounding trumpets and other refinements.

If, as in Athenian law, each side in a law-suit was allowed only a specific amount of time for speaking, this was measured with the clepsydra. The phrase 'in my water' meant 'in my allotted time'. The clepsydra came to Rome in 159 BC and from then on was used as well as the sundial; it was particularly needed in the law-courts, and in army camps it served to demarcate the night watches. The octagonal Tower of the Winds in Athens, erected in the first century BC, had an accurate clepsydra, of which there are remains, and a series of planar sundials on different faces. A water-clock described by Hero of Alexandria included a graduated drum whose lines indicated the hours at any season.

Astronomical Calendars

In the fourth and third centuries BC, Greek astronomers devised elaborate models to illustrate the motions of heavenly bodies. One such, invented by Archimedes, was taken to Italy after the capture of his city, Syracuse, in 212 BC. Another of the same type has been called 'an ancient Greek computer'. Perhaps more realistically it may be called an astronomical calendar. In 1900 the captain of a Greek sponge vessel rescued statues and other objects from a wrecked Roman ship, off the island of Anticythera, between the Peloponnese and Crete. The following year a scientific instrument was found there. It contained at least 20 gear wheels, and had two dials, one recording the apparent motions of the sun and stars, the other, evidently, those of the moon and planets. Presumably a slave would move the handle at regular intervals, since, although this instrument contained clockwork, it was not a clock. The last setting marked was in 82 BC, and the statues and this instrument were clearly intended to be transported to Italy as 'spoils of war' after the sack of Athens in 86 BC.

8 Calculations for Trade and Commerce

Early Trade

Trade existed in the ancient world long before the invention of coinage. Goods or services could be exchanged at rates agreed by the parties involved in the transaction. Promissory notes could also have been used. Such a note might be accepted if, for example, a harvest was not ready at the time when the transaction was due to take place.

Gradually a system was developed to facilitate trade and the making of payments in general by assessing values in the form of weighed amounts of metal. One of the first attempts at standardisation was probably that of the Sumerians, who from *c.* 2400 BC inscribed basalt statuettes of sleeping ducks with their correct weights.

43 Stone 2-talent weight in the shape of a duck, from Lagash, Mesopotamia, *c.* 2260 BC. British Museum.

Main Sumerian weights

The Sumerian weight system was based on sexagesimal notation:

60 shekels, of approx. 8 g each = 1 *mina*, approx. 480 g
60 *minae* = 1 talent, approx. 28.8 kg

Shekels and talents are not the original Sumerian names: these and other equivalents may be tabulated thus:

Sumerian	Babylonian / Assyrian	English
gú	biltu (literally 'load')	talent
ma-na	manû	mina
gín	šiqlu	shekel
še	uṭṭetu	grain
anše	imēru (literally 'donkey')	homer

It will be seen that 'shekel' is a Hebrew form of *šiqlu*; '*mina*' and 'talent' are adapted through Latin from Greek forms. We find considerable variations from these Sumerian weight values elsewhere, from 25 shekels to a *mina* in Phoenicia to 100 in Egypt. The weight of the shekel also varied.

Egyptian weights

In Egypt, from the Eighteenth Dynasty onwards, metals were weighed as follows:

10 *kitĕ* (approx. 9.1 g) = 1 *deben* (approx. 91 g)

44 Egyptian stone weight inscribed with the name of King Amenophis I and the value 5. Eighteenth Dynasty, *c.* 1550 BC. British Museum.

45 An Egyptian weighing gold rings against a weight in the form of a bull's head. New Kingdom, *c*. 1400 BC. British Museum.

Main Greek weights

	Aegina	Attica / Euboea
obolos	1.04 g	0.73 g
6 *oboloi* = 1 *drachmē*	6.24 g	4.36 g
100 *drachmai* = 1 *mna*	624.00 g	436.00 g
60 *mnai* = 1 talent	37.44 kg	26.196 kg

Roman weights

1 *uncia* = 27–27.5 g
12 *unciae* = 1 *libra* (pound)

The English abbreviation 'lb', the £ sign and the word 'pound' all have Latin antecedents, since £ and lb = *libra* and pound = *pondo*, 'in weight'; in Latin, P. was used as an abbreviation for (*librae*) *pondo*.

2 *unciae* = 1 *sextans*, $\frac{1}{6}$ *libra*
3 *unciae* = 1 *quadrans*, $\frac{1}{4}$ *libra*
4 *unciae* = 1 *triens*, $\frac{1}{3}$ *libra*
5 *unciae* = 1 *quincunx*
6 *unciae* = 1 *semis*, $\frac{1}{2}$ *libra*
7 *unciae* = 1 *septunx*
8 *unciae* = 1 *bes*, $\frac{2}{3}$ *libra* (cf. *bis*, 'twice')
9 *unciae* = 1 *dodrans*, $\frac{3}{4}$ *libra* (literally '1 *quadrans* short')
10 *unciae* = 1 *dextans*, $\frac{5}{6}$ *libra* (literally '1 *sextans* short')
11 *unciae* = 1 *deunx* (literally '1 *uncia* short')

There were also subdivisions of the *uncia*: the *siculus* ($\frac{1}{4}$), the *sextula* ($\frac{1}{6}$), and the *scripulum* ($\frac{1}{24}$).

48

Three types of weighing-machine were used in the ancient world. If the object was small, it was convenient to use scales with a pivoted beam. In some cases the weights for these were of different shapes, in other cases of the same shape, but they were always graded. The other two kinds of ancient weighing-machine, useful for large items, were the bismar and the steelyard. The bismar is mentioned by Aristotle (384–322 BC) as serving to weigh large quantities of meat. It had a fixed counterpoise at one end of a beam, at the other a hinged hook to carry the meat. Attached to the beam was an upper bar with graduated lines to indicate the weight. This upper bar was slotted into a vertical holder and was pushed along until the bismar reached equilibrium. Although this type survived until the Middle Ages, it was inaccurate and was banned from churches in England in 1428. The Roman steelyard (*statera*) is described by Vitruvius. This, unlike the bismar, has a fixed fulcrum and a movable counterpoise. An improved model was introduced in the fourth century AD.

Since weights and measures often differed from area to area, and since scales were not always reliable, it is not surprising that cheating occurred. The prophet Ezekiel (45.9–12) warns: 'Enough, princes of Israel! ... Your scales shall be honest, your bushel [*ephah*] and your gallon [*bath*] shall be honest. There shall be one standard for each, taking each as the tenth of a homer, and the homer shall have its fixed standard. Your shekel weight shall contain 20 *gerahs*.' After this the text is corrupt, but the latest editor, W. Zimmerli, relying on the Septuagint, renders: '5 shekels are to be 5, and 10 shekels are to be 10, and 50 shekels are to amount to a *mina* with you.'

Coinage

Once coinage was invented it was not long before it became the usual medium of exchange for trade in many areas of the ancient world. The date and identification of the very first coins is disputed, but by the end of the seventh century BC coinage was circulating in Lydia, western Asia Minor, whose last king, Croesus (*c.* 560–546 BC), is proverbial for his wealth. The earliest coins were of electrum (white gold) and were one-sided; coins with fully developed designs on both sides came later, towards the end of the sixth century BC.

In the classical period the words used for monetary terms of account mostly reflect their association with weight, the name *mna* or *mina* in particular being derived from the name of the Mesopotamian weight unit given above. The names of the small Greek weights and coins *obolos* and *drachmē*, the latter still the unit of Greek currency, literally meant 'metal spit' and 'handful'. The Latin word for money, *pecunia*, was derived from *pecus*, 'cattle', reflecting the pastoral interests of early Rome.

Coinage was normally issued by states, but it was sometimes signed by individuals, usually working for the state, though sometimes perhaps as private persons. An early electrum coin, probably from Halicarnassus in Asia Minor, reads *Phanōs ēmi sēma*, 'I am the badge of Phanes'. Coins were not necessarily uniform in size, but were supposed to be so in weight, and the verbal association already noted helped this; the weighing of coins in transactions therefore continued. Most coins were made of gold or silver and care had to be taken to ensure that they were of good metal as well as the proper weight.

46 Athenian didrachm. The obverse shows the goddess Athena, patron of Athens, and the reverse an owl, symbol of Athena and badge of the city. British Museum.

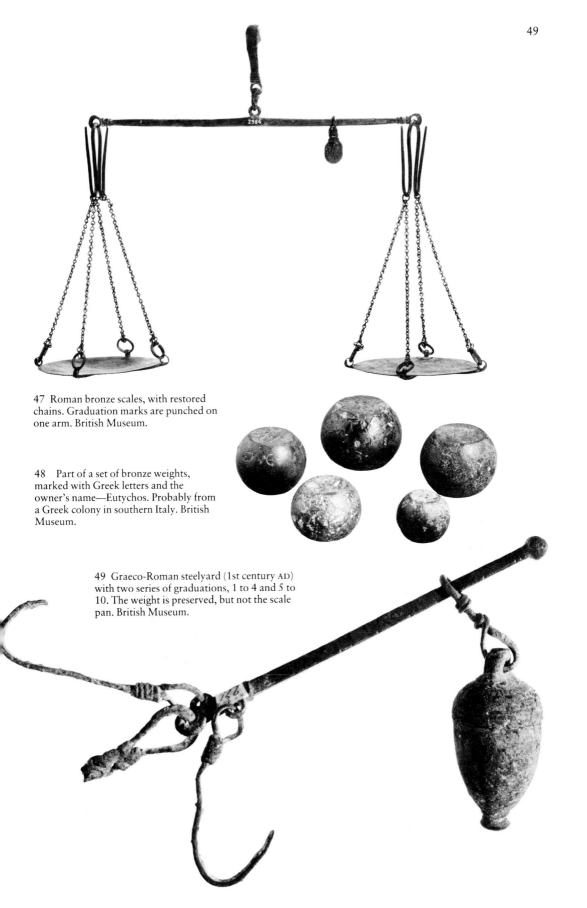

47 Roman bronze scales, with restored chains. Graduation marks are punched on one arm. British Museum.

48 Part of a set of bronze weights, marked with Greek letters and the owner's name—Eutychos. Probably from a Greek colony in southern Italy. British Museum.

49 Graeco-Roman steelyard (1st century AD) with two series of graduations, 1 to 4 and 5 to 10. The weight is preserved, but not the scale pan. British Museum.

Main Greek coinage
This was based on the *obolos* and *drachmē*, as in the table of weights above, together with the *didrachmon* or *statēr*, equal to 2 *drachmai*. In Athens after 530 BC the *didrachmon* was replaced by the *tetradrachmon* (4 *drachmai*). There were many local variations from time to time, summarised in *The Oxford Classical Dictionary* under 'Coinage, Greek'. For the higher terms of account, see the weights table above.

Roman coinage

> 12 *unciae* = 1 *as*, with the same functions as for weights
> $2\frac{1}{2}$ (later 4) *asses* = 1 *sestertius*, abbreviated IIS ($2\frac{1}{2}$) or HS
> 4 *sestertii* = 1 *denarius* ('containing 10', since originally $2\frac{1}{2} \times 4 = 10$; abbreviated to *d.* for the old British penny)
> 25 *denarii* = 1 *aureus*

Although the *as* was originally equal to the *libra*, it became successively devalued, and other coinage with it; thus in Caracalla's reign (AD 212–17) there were 50 *aurei* to a *libra*. Very large sums were normally reckoned in thousands of *sestertii*; from this, the genitive plural, as in *duo milia sestertium* (2000 *sestertii*), came to be looked on as a neuter noun, so that we also find *duo sestertia*.

From the point of view of trade, it helped the customer that subdivisions of the *as* were the same whether they referred to coinage or weight.

Since every citizen and any educated slave was expected to acquire a complete mastery of the monetary system, it came to occupy what some considered a disproportionate amount of time in the school curriculum. Horace, in the *Ars Poetica*, complains that this is a wearisome part of Roman education:

> Children at Rome learn to divide the *as*
> By lengthy workings into tiny parts.
> 'Son of Albanus, tell me: if we take
> An *uncia* from a *quincunx*, what is left?
> Be good enough to say.' 'A *triens*.' 'Fine!
> Your money will be guaranteed for life.
> We add an *uncia*: what does that make?' 'Half.'

In other words, $\frac{5}{12} - \frac{1}{12} = \frac{1}{3}$ *as*; $\frac{5}{12} + \frac{1}{12} = \frac{1}{2}$ *as*. Although in the second line the Latin says *centum partes*, '100 parts', this is only poetic licence, since the initial division of the *as* was into 12 *unciae* and, even though these could be further subdivided, $\frac{1}{100}$ would never be a relevant fraction of the *as*.

Accounting and Banking
Financial lists and accounts have survived from antiquity, and better on Egyptian papyri than on other materials elsewhere. The tribute lists of the Athenian treasury, recording Athena's share of the tribute paid by allies to the Delian League (478–447 BC), are of more interest to political historians than to students of the history of mathematics and measurement. Papyrus finds include public and private accounts, estate and taxation documents, and military documents, some of these of a statistical nature.

Certain banking operations are known from very early times; but the earliest known equivalents of private banks, linked with estate management, are Mesopotamian of the seventh century BC. Greek banking was carried out by individual bankers, who had no offices, only tables in the central area of the city. Hence *trapeza*, 'table', could from before 500 BC mean a bank, and indeed does so in modern Greek. Throughout the ancient world, although various religions and states denounced usury, interest was regularly charged, and debtors could be thrown into prison. But the death or default of a banker could create serious problems.

This situation was temporarily rectified by the procedure in Ptolemaic Egypt. A central bank was set up in Alexandria, with royal banks in many other towns, lending money as well as receiving deposits, and employing many bank clerks. Unfortunately this precedent was not followed up under the early Roman Empire, except in Egypt. It was private bankers who, in Greece on the last day of the month and in the Roman world on the Kalends of each month, demanded interest payments, which were calculated monthly, not annually. Unpaid interest might be added to the capital, so as to be treated as compound interest. Contrary to the general custom, under the late Empire a branch of the civil service concerned itself with many aspects of banking.

Roman taxation

Taxes in the Roman Empire were of two basic types, *vectigalia* and tribute. The main *vectigalia*, of which the first two were already enforced under the Republic, were:

Name	Explanation	Percentage
portoria, usually *quadragesima* ($\frac{1}{40} = 2\frac{1}{2}\%$)	Customs duty, originally at ports, later also at land frontiers	$2\frac{1}{2}\%$ (higher in the Near East)
vicesima libertatis	Tax on value of slaves freed	5%
quinta et vicesima venalium mancipiorum	Tax on value of slaves sold	4%
centesima rerum venalium	Tax on auction sales	1% (Augustus), $\frac{1}{2}\%$ (Tiberius)
vicesima hereditatum	Legacy duty; close relatives exempted	5%

In addition, *vectigal* was levied on occupiers of State lands, *ager publicus*. The second type of taxation was tribute, not paid by Roman citizens under the Empire. In at least two provinces the rate of this was 1% per annum on the *census*, total personal capital including houses, land, slaves and other possessions.

Greek interest rates

Expressed as 'at 1 drachma', 'at 8 obols', etc., meaning 1 drachma a month or 8 obols a month for every *mina* borrowed, equivalent to 12% or 16% per annum.

Roman interest rates

Also reckoned monthly:

triente = $\frac{1}{3}\%$ per month = 4% per annum

besse or *bessibus* = $\frac{2}{3}\%$ per month = 8% per annum

centesimis = 1% per month = 12% per annum, the rate laid down by the Senate in 51 BC

binis centesimis = 2% per month = 24% per annum

All these Latin forms are ablatives of price, *triente* from *triens*, *besse* from *bes*, the others from *centesimus*, $\frac{1}{100}$. Inheritance proportions were reckoned in the same way; thus *ex triente* = $\frac{1}{3}$ of the estate, *ex quadrante* = $\frac{1}{4}$ of it.

A Specimen Trade

The use of mathematics and measurement in trade may be conjectured for a business such as that of a timber merchant in Italy. First he would need to calculate the volume of each type of timber ordered, taking account of local measurements in which it would be quoted. Then he had to negotiate a price, working out the foreign exchange if it was to be imported. After this we may suppose that he agreed to costs of land or sea transport, if necessary consulting an itinerary for the distances involved. We know from Diocletian's price edict of AD 301 that in his time land transport of heavy goods was much more expensive than sea transport. In the case of the latter, however, it was customary to pay for bottomry, an approximation to shipping insurance. A moneylender could make a loan to a shipowner, which did not have to be repaid if the ship and cargo were wrecked or seized by pirates. At the land frontier or harbour a percentage tax was levied. For sawing or planing, more slaves might have to be bought, at a higher rate if the work was skilled. If exact measurements of timber were specified, a measuring-board or slab would be used. Such a stone slab, found in Italy, is in the reserve collection of the British Museum. It has no inscriptions, but to judge from the frequency of 3-*digiti* (finger's-breadth) marks, that measurement must have been a common one for the user, who also worked in feet and might have been engaged in selling cloth, leather straps, timber or marble. When all the expenses arising from the timber purchase had been evaluated, the merchant would be able to fix a price for each type and size. To handle complexities of foreign trade, a Greek-speaking freedman would most often have been the best qualified.

9 Mathematics in Leisure Pursuits and the Occult

Puzzles

Among the leisure interests of the past was the simpler sort of arithmetical puzzle. Of the 'think of a number' type, our best specimens are from the Dark Ages, but clearly reflect earlier usage. They are preserved in the treatise *De arithmeticis propositionibus*, wrongly ascribed to the Venerable Bede. Exercise 1 may be translated: 'Let any number be thought of and trebled. Divide the trebled number in two, and if this results in equal integers, again treble one of these. If they are unequal, let the greater of the two be trebled. Record the number of times 9 can be divided into this, then twice that number is the one thought of. But if there is a remainder, which will be 6, add 1 of this 6 to the total reached above.' The exercise may easily be proved to be sound. There are unequal parts when there is an odd number. Thus, think of 11; treble it, 33; divide by two, 16 + 17; treble the larger number, $51 = (9 \times 5) + 6$; $5 + 5 + 1 = 11$. Another type of puzzle consisted of 14 ivory shapes which could be put together to resemble animals.

The invention of chess in India or possibly China is associated with a story about geometrical progression. An Indian king, very pleased with the inventor, despite the limitations placed on the powers of the chess king, asked what favour he could do him. The chessplayer asked for 1 grain of wheat for the first square, 2 for the second, 4 for the third and so on. The king granted this, not realising that the total for 64 squares came to $2^{64} - 1$, or 18,446,744,073,709,551,615 grains. This was reckoned to be equivalent to eight harvests all over the world.

Games

Whereas chess did not reach the Mediterranean till after classical times, something resembling draughts seems to have been played in dynastic Egypt. Pictures show the pieces set up in three rows on a board, just like a modern draughts-board. No particular mathematical

50 A gazelle playing a board game with a lion. Part of a painted papyrus from Egypt. New Kingdom, *c*. 1200 BC. British Museum.

51 Graeco-Roman dice. British Museum.

skill was involved. Several ancient games, which were and still are played in many parts of the Near East and Mediterranean countries, involved simple mathematics. One such is the guessing of the number of fingers the opponent will hold up. The usual method is for two people simultaneously to hold up between 1 and 4 fingers: the first to guess the combined total wins. For a medieval number game, see Smith (1923), 198–200.

Dice in the Graeco-Roman world were marked as today, i.e. with opposite faces adding up to 7, and they sometimes occur in left-handed and right-handed pairs. The lowest throw, 1 on each die, was called in Latin *canis*, 'dog'. They were thrown either in a dice-box or in a model of a tower. Despite legal prohibitions at various times, gambling, which also involved calculations, remained a regular accompaniment of dice-playing and chariot-racing. Dice-playing often involved board games with counters, for which a certain element of skill as well as luck was needed.

Among these, two Roman games, which may have had Greek or Near Eastern antecedents, can be reconstructed with some probability from finds and literary sources. The first was called *latrunculi*, 'little mercenaries'. Since one of the Piso family—probably C. Calpurnius Piso (died AD 65), leader of the unsuccessful conspiracy against Nero—was very fond of it, there is a poetic description of it in the anonymous *Laus Pisonis*, though in that there is no mention of the name *latrunculi* or *latrones*. It was played on a board with white and black or red pieces, which are variously called *milites* (soldiers) or *latrones* (early name for mercenaries, later for bandits). The board is conjectured to have been of 8 × 7 squares, such as was found at Corbridge, Northumberland. A piece was taken if it was surrounded on rank or file by two opposing pieces. Backward as well as forward moves were allowed, and one line of the verse suggests a rook's move in chess. A blocked piece could be extricated by a skilful player. The game was won by the player who removed more pieces; he was hailed *imperator*.

52 Roman gaming board, inscribed with a sentence of six six-letter words: CIRCVS PLENVS CLAMOR INGENS IANVAE TE(NSAE) ('circus full, terrific shouting, doors bursting'). British Museum.

The other game was called *duodecim scripta*, 'twelve writings'. Although some scholars have disputed the association and speak of a 36-letter game, distinct from the 'twelve writings', one cannot help interpreting *scripta* here as sentences of 6×6 letters, each group of six letters forming one or two words. In the Latin Anthology there is a set of 12 such hexameter sentences, which may suggest that a complete game set consisted of twelve six-word, six-letter sentences, together with three dice, one dice-box, probably 15 white and 15 black or red counters, and a board with 3×12 intersections, the 12 being split 6 and 6. An example of a sentence in the British Museum (fig. 52) may be rendered: 'Circus full, terrific shouting, doors [?] bursting'. Some examples have a moral tone on gaming, such as the hexameter of six-letter words which reads IRASCI VICTOS MINIME PLACET, OPTIME FRATER: 'dear brother, I can't stand losers who lose their temper.' But the tone was not always moral, since some are about enjoying life.

What role these sentences played in the game is not clear. One inscription, perhaps intended for beginners, reads:

CCCCCC	BBBBBB
AAAAAA	AAAAAA
DDDDDD	EEEEEE

The order of the letters is presumably a guide to the direction of moves. It implies that the pieces were first placed on the A squares, each according to the number on a single die throw; then, when all had reached an A square, they could progress via the second set of A squares, B, C, D, E; though some maintain that black took these last four in reverse order. It seems that three six-sided dice were normally thrown, and their numbers used to move one, two or three pieces. The maximum throw was $3 \times 6 = 18$, and point XIX (presumably the first C reached), which a counter could reach from I in one turn, was called *summus*, 'top'. If a white piece landed on a point containing one black piece, it captured it, probably sending it back to point I, and *vice versa*. But if there were two or more enemy pieces on a square, the proposed move failed. Hence the strategy was to try to avoid having unaccompanied pieces. In fact this aspect of the game is very similar to the rules of backgammon. The game was won when all the pieces of one side had reached the opposite end. In a wall-painting from Pompeii, one player says *exsi*, meaning *exii*, 'I've got through', while the other says *non tria, duas est*, 'It's not a three, it's a two.'

Betting was common, but only in one passage of Plautus do we find mention of betting for odds: 'Bet me a talent to a drachma [or similar small coin] ...' In view of the absurdly high odds (well over 1000 to 1), this must be comic exaggeration rather than reflect regular practice.

The Occult
The connection between mathematics and the occult in antiquity is widespread. Certain numerals either acquired occult meanings or were regarded as particularly auspicious or inauspicious. The origin of the Jewish sevenfold candlestick goes back to words spoken by God to Moses (*Numbers* 8.2): 'Speak unto Aaron, and say unto him, "When thou lightest the lamps, the seven lamps shall give light over against the candlesticks."' There are magic numbers in many religions: three is often such a number. A magic charm may have to be repeated three times, or three deities may be worshipped. The great temple on the Capitol at Rome was dedicated to Jupiter, Juno and Minerva; some Roman colonies, instead of one large temple, had three small ones on their Capitol. Four was Hermes' number, seven was Apollo's, nine (3×3) was the number of the Muses. Numerals may also represent letters and *vice versa*. One such example is the sundial with successive letters ZHΘI, either 'live' (long may you live) or 7, 8, 9, 10, four successive hours on the dial (fig. 53). A better-known example is in the *Revelation of St John the Divine*. There 666 represents

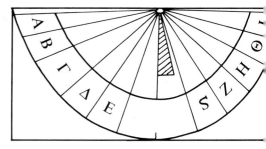

53 An ancient epigram (*Anthologia Palatina*, x.43) may be rendered 'Six hours of work are quite enough to give: the next four stand for ZHΘI, meaning "live".' This drawing of a sundial gives hours 1–10 as Greek letters, including 7–10 as ZHΘI, 'live', i.e. 'live long'. It was first published in the *Literary Gazette*, May 1823, purporting to show a sundial found at Herculaneum, though one suspects it may have been reconstructed from the epigram and from the sundial in fig. 40 or one similar.

by the Hebrew numerical value of its letters a man's name, so the Greek text as usually now interpreted tells us, and we have to work out what this is: one answer suggested is the first emperor who persecuted Christians, Nero. This 'mark of the beast' occurs at the end of Chapter 13. Earlier religions had given a mystic connotation to 13 as being one more than 12, a round dozen used among other things for the number of the tribes of Israel. When under Christianity 13 came to be associated with the Last Supper, it acquired an inauspicious meaning, but we do not know if the chapters of this book were marked off at an early date.

It is not surprising that echoes of these associations can be found in classical poetry. Virgil, who studied mathematics in his youth, seems to have been fond of sentences or groups of sentences occupying exactly 3, 7, 9 or 12 hexameter lines. As an oddity, Horace's Ode i.11, *tu ne quaesieris*, in which he prophesies that he and his patron Maecenas will die about the same time, contains 56 words. The prophecy proved correct, as Maecenas died within weeks of Horace. But the poet may also have been prophesying by a riddle that he would be 56 at his death: in fact he was 56 and 11 months.

Astrology and clairvoyance were prevalent in most periods of antiquity. Not only were the twelve signs of the Zodiac used, but complete horoscopes could be drawn up. One such is the pictorial inscription known as the Lion of Commagene. Commagene was a small kingdom in the north of Syria, and the horoscope of its king, Antiochus, is preserved in a bas-relief formerly dated to 98 BC but now held to be of 62 or 55 BC. The king was born in the sign of the Lion, so a lion is shown with stars outlining parts of its body, and with three planets above it.

Finally there is what may be called the 'intentionally occult', the use of codes and ciphers. Unfortunately we know little of this in antiquity except that Julius Caesar used a simple progressive letter substitution for his despatches to Rome from Gaul; but there may have been some cryptography which could be called arithmetical.

54 The Lion of Commagene, a bas-relief depicting the horoscope of Antiochus I, king of Commagene *c.* 70–38 BC. Nimrud-Dagh, south-east Turkey.

10 The Sequel

As the research by Graeco-Roman mathematicians diminished about AD 500, that of Indian mathematicians, who had contacts with the West, increased. Among their accomplishments were the extraction of square roots; the best approximation of π up to that time, 3.1416; an estimate, by astronomical methods, of the distance from central India to 'the Greek city', evidently Alexandria; and predictions of the future positions of planets.

An aid to their calculations was a simplified form of numerals. It is true that Greek and Roman numerals have in part survived to the present day: the former are used in the alphabetic form to denote volume numbers, etc.—e.g., in modern Greek Vol. 2 is Tom. B; the latter are used particularly to express not only chapters or volumes but years, as MCMLXXXVI for 1986. But both these and their Near East predecessors had distinct disadvantages. Hindu numerals were passed on to the Arabs, via Baghdad, about AD 800, and we know them as arabic numerals, though in fact our forms of 2, 3 and 5 are slightly nearer to Hindu than to the normal Arabic forms. Probably 2 and 3 originated as = and ≡ joined up, though in Arabic these strokes are vertical. An important part of this notation was the incorporation of a sign for zero. Our sign 0 seems to have arisen as a Greek astronomical zero co-ordinate, for example in the *Almagest* of Ptolemy, though there was no zero in other types of Graeco-Roman mathematics. In Hindu notation it decreased in size, and in Arabic numerals proper, where 0 signifies 5, zero became a dot. Comparative tables for 0–2 are:

Late Greek	Hindu	Arabic	Spanish	Italian	Modern
O	o	·	O	o	0
α	؟	١	I	I	1
β	◌	٢	ح	2	2

The use of numerals with plus or minus signs is found in various forms in the Chinese treatise *Nine Chapters on the Mathematical Art* (50 BC or later), in Diophantus (for minus signs only), in the Indian Brahmagupta, and in pseudo-Bede's treatise. It was followed much later by signs for multiplication and division. With these aids written arithmetical calculations became much easier. By the fifteenth century arabic numerals, as we call them, though earlier frowned upon, were used extensively for trade and navigation; they had long been incorporated into mathematical treatises.

The Arabic centre which promoted most of the sciences was Baghdad. Under Haroun al-Rashid (c. 763–809), Caliph of Baghdad, scholars were gathered from every civilised country of the Near East and the eastern Mediterranean, and material brought by them and by merchants was collated. As a result, a number of Greek mathematical and technical works were preserved in translation. The latitudes and longitudes of Ptolemy's *Geography* were revised to take account of more recent findings and to include Arab settlements which did not exist in Ptolemy's time. Al-Idrisi, who worked in the period up to 1154 for the Norman king Roger II of Sicily, and other Arab cartographers produced world and regional maps (fig. 55). Indian and Arab sundials and clocks and Arab water-wheels imitated and often improved on classical models. With the establishment of a seat of learning at Cordova, European scholars, such as Abbot Gerbert (later Pope Silvester II) in the tenth century, went there to meet Arab scholars and find out what scientific works had been preserved from the Greek world. Gerard of Cremona, working in Toledo in the twelfth century, translated Arabic versions of Euclid, Archimedes and others.

55 World map by the Arab geographer Al-Idrisi, with south at the top. The original was commissioned by the Norman king Roger II of Sicily and completed in 1154. Oxford, New Bodleian Library, Ms Pococke 375, f.3v–4r.

During the Dark and Middle Ages, Western European measurements varied from country to country (or even within countries) and only in part reflected Roman practice. Thus in England, the foot (30.48 cm) was longer than the standard Roman foot (approx. 29.6 cm); instead of 5 ft = 1 (double) pace, the next unit is the yard of 3 ft, followed by the chain of 22 yd and the furlong of 220 yd; and the mile is 5280 ft, not 5000 as in Roman measure, and is equal to 1.6093 km, instead of about 1.48 km. For a long time a descendant of the Roman *ulna*, 'forearm', was in use in various countries. The English ell measured 45 in, which bears no relation to the length of the forearm, as the derivation of the word would suggest.

There is even a legacy of Roman land survey which has survived to the twentieth century. In Roman times the surveyor's pole was called *pertica* or *decempeda*, and its length, as implied by the latter name, was 10 Roman feet (approx. 2 m 96 cm). The English measure was 'rod, pole or perch', and this last name is derived from *pertica*. But it was normally $5\frac{1}{2}$ yards (5 m 3 cm) long, and this is why there are 4840 square yards to the acre. In earlier times English, Irish and Scottish acres all measured 160 square poles, but the pole varied considerably, the English acre being $5\frac{1}{2} \times 5\frac{1}{2} \times 160$ yd². In fact the measurement of $5\frac{1}{2}$ yards dates from about 1305, a Norman English equivalent being 16 feet, not $16\frac{1}{2}$.

Arithmetic was simplified by the introduction of arabic numerals. As an example of complicated calculations involving arithmetic and geometry, Al-Kashi, who about AD 1400 was the first director of the Samarkand observatory, wrote a treatise on the value of π. He followed the lead of Archimedes, but on a gigantic scale, based on inscribed and circumscribed polygons with an extremely large number of sides. But advances in arithmetic and geometry in various parts of the Old World had been so spectacular that we need not be surprised if trigonometry and algebra received more attention in Renaissance research.

Some elements of trigonometry were established early in the history of mathematics, and the Greek mathematician Menelaus (*c.* AD 100) enabled spherical trigonometry to be applied to astronomical plotting. Further advances, as seen above, were made by Indian mathematicians after the fall of the western Roman Empire; while the sine theorem, known from about AD 1000, was attributed to the Arabic writer Abu al-Wafu. But it was left to a sixteenth-century German, influenced by a thirteenth-century Persian, to codify this branch of mathematics. Regiomontanus, or Johann Müller of Königsberg, who went to Leipzig University at the age of 11, may be called the father of modern trigonometry. He was influenced by Nasir al-Din al-Tusi (1201–74), a Persian who wrote in Arabic. Regiomontanus' work *De triangulis omnimodis* (1533) caused trigonometry to be looked on as a discipline of its own. His name for sine was *sinus totus* and for cosine *sinus complementi*; this was first called *co-sinus* in 1620.

Algebra had been developed as a logical system in the later Greek world of Alexandrian learning by Diophantus. Indian mathematicians included in their research Diophantine equations, those with two or more variables, of which solutions are required to be integers. Greater advances were made from the sixteenth century. In 1545 the Italian mathematician Girolamo Cardano, the title of whose book may be rendered as *The great art, or On algebraic rules*, was the first to publish the solution of cubic equations, perhaps discovered earlier in that century by other Italians.

The application of mathematics and measurement to Renaissance technology depended to no small extent on the legacy of the Graeco-Roman world. Small-scale mapping, surveying, architecture, timekeeping and astronomy were among the fields particularly affected.

By the late Byzantine Empire the *Geography* of Ptolemy had become neglected, and when in 1295 Maximus Planudes found a copy in Constantinople it lacked maps. However, he was soon able to recopy the text with world and regional maps, either copied with improvements from another existing manuscript or reconstructed from the text. The major breakthrough came in 1406, when Jacopo d'Angelo in Florence translated it into Latin. Since for many parts of the world Ptolemy's maps improved on or complemented existing maps, the *Geography* was first widely duplicated in manuscript and then circulated in printed form—from 1475 without maps and from 1477 with them. In many of these printed editions 'modern' or 'new' maps appeared, particularly of parts of the world not known to the ancients. Although, coupled with other sources, editions of Ptolemy's *Geography* misled

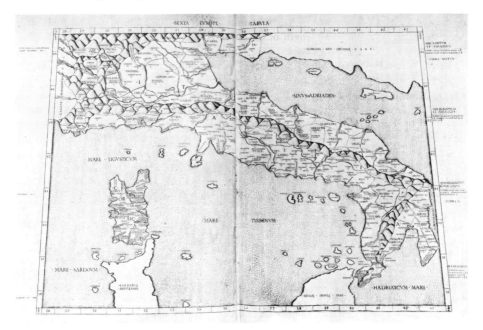

56 Map of Italy from Ptolemy's *Geography*, Rome 1478 edition. On this map the east coast takes an abrupt turn south instead of continuing ESE. All manuscript variants of Ptolemy's latitude and longitude co-ordinates make this appearance of south-east Italy inevitable, with the 'heel' south of Monte Gargano, not south-east. Ptolemy's co-ordinates further north had already gone too far east. British Library.

Columbus into thinking that in the Caribbean he had reached Asia, they played an outstanding part in the history of exploration.

Progress in surveying technique was slow until the seventeenth century. The Romans had concentrated on land survey by squares or rectangles on the whole, but by Tudor times graphic triangulation was becoming important, with the introduction of plane-tabling. Only by the seventeenth century were the Greek *chorobates* and *dioptra* and the Roman *groma* and measuring-rod superseded. It was as late as 1700 that spirit-levels took the place of the *chorobates* and its successors. The *dioptra* was totally ignored, but its place was taken first by instruments based on trigonometrical advances and logarithmic tables, facilitating angle-measurement in the field, and ultimately by the theodolite. The cross-staff, successor to the *groma*, was itself replaced by the incorporation into the theodolite of the vernier, invented by the Frenchman P. Vernier, and telescopic sights; while measuring-rods were bettered by Edmund Gunter's development of a new surveying chain.

In the regular planning of the countryside we cannot be so sure that Roman antecedents played a part; if they did, it was not from examination of the remains (no work was done on this until 1833), but through circulation of Willem Goes' 1674 edition of the Roman land surveyors' manuals. It is likely that Thomas Jefferson consulted these before devising in 1784 his own schemes for land division of the United States mid-West.

Vitruvius' *De Architectura*, of which many manuscripts exist, was printed in Rome about 1486, in Florence in 1522, and in Rome again in 1544, and the work came into circulation in Italy, where hundreds of large Renaissance buildings still survive. It particularly influenced the Italian architect Andrea Palladio (1508–80), who showed his originality by adapting rather than copying Graeco-Roman buildings. Thus the Teatro Olimpico at Vicenza (fig. 57), started in the year of his death and completed three years later, is based on a Roman theatre, but has a permanent ceiling over the auditorium, the first theatre to have this feature, and ingeniously planned passages receding into the distance and creating a sense of perspective for the audience. The auditorium itself is not semicircular but forms a segment of a circle.

Among Palladio's other conspicuous buildings is the so-called Rotonda, or Villa Capra, in the countryside near Vicenza (fig. 58). This was built between 1550 and 1553 as a fanciful interpretation by Palladio of villas of Cicero and the younger Pliny. It is a perfectly symmetrical domed building with four identical Ionic façades, so that the villa, as Palladio writes, 'enjoys on all sides most beautiful views'. The entire neo-classical architectural revival owed much to Palladio's study of Vitruvius, and measurement played a significant role in this revival.

In timekeeping, the Middle Ages showed an improvement on the ancient methods of water-clocks and hour-glasses. Clocks driven by heavy weights were installed in churches.

57 Palladio's Teatro Olimpico, Vicenza: interior.

58 Palladio's Villa Capra, or Rotonda, near Vicenza.

Among famous specimens is that in Strasbourg Cathedral, built by Henri de Vick (Wieck) in 1362–70 and operated by an escapement and a weight. The first important step forward, the invention of the pendulum clock, came after Galileo in 1583 discovered that a wider swing of the Pisa Cathedral chandelier took no longer than a narrower swing. Later, by checking with a water-clock, he found that the length of a pendulum varies as the square of the duration of its swing. This discovery, which strictly applies only to narrow swings, enabled far more accurate clocks to be designed.

During the Dark and Middle Ages the Julian calendar, supplemented by the seven-day week introduced in the fourth century AD, continued unchanged. But as it was based on a year averaging $365\frac{1}{4}$ days, whereas the true value is 365.242199, there was an error after 1000 years of 7.801 days. For centuries, despite pleas, nothing was done to rectify this discrepancy. But in 1545 the Council of Trent made positive recommendations, which a generation later resulted in the first substantial change since Julius Caesar. In 1572 Pope Gregory XIII appointed a Jesuit astronomer to formulate a new calendar, and ten years later this Gregorian calendar was adopted in many European countries. It provided that ten days should be omitted that year, thus incorporating a slight correction to bring the spring equinox back to 21 March, and that centennial years (1600, 1700, etc.) should not be leap years; this itself was subsequently amended so that years divisible by 400 should be leap years. In 1752 Britain adopted this 'New Style', omitting eleven days.

Advances in astronomy came in Muslim areas, with observations at Istanbul, Samarkand and elsewhere, and then in Europe in the mid-sixteenth century, when Copernicus turned away from the geocentric theory of nearly all previous cosmologists. Among Greek scientists, Aristarchus (third century BC) had maintained that the earth and planets go round the sun. This theory was confirmed during the century after Copernicus, when Kepler showed that the orbit of each planet is not a circle but an ellipse. As a result, ingenious theories by Greek astronomers to explain apparent orbit discrepancies could for the first time be disproved. The ancients, for example, had no idea of the size of the sun or the distance between the sun and the earth (though for the moon Hipparchus came close to the right answer). It was the invention of the telescope and observations by Galileo, who made the first astronomical telescope, that led to rapid advances in astronomical measurement. But it was not until the seventeenth century that the distance of the sun was able to be calculated within about 10%.

To summarise, in the period between the fall of the Roman Empire and the Renaissance we need to look at mathematics and measurement separately. Measurement, with the dividing up of the western provinces, came to have many more regional differences, even within countries; and the persistence of Latin as a scholarly language was not accompanied by an adherence to Roman weights and measures. Research in mathematics, on the other hand, first pure and then applied, continued over a wide area of the Old World and, with the rediscovery of classical accomplishments, led to remarkable advances during the Renaissance.

Bibliography

General

Bernal, J. D. *Science in History*. Vol. 1: *The Emergence of Science*. 3rd edn, Harmondsworth, 1965.
Berriman, A. E. *Historical Metrology: A new analysis of the archaeological and the historical evidence relating to weights and measures*. London and New York, 1953.
Bickerman, E. J. *Chronology of the Ancient World*. London, 1968.
Boyer, C. B. *A History of Mathematics*. New York, 1968.
Eves, H. *An Introduction to the History of Mathematics*. 2nd edn, New York, 1964.
Harley, J. B. and Woodward, David. *The History of Cartography*, Vol. I. Chicago, 1987.
Hodges, Henry. *Technology in the Ancient World*. London, 1970.
Hogben, Lancelot. *Man Must Measure: the Wonderful World of Mathematics*. London, 1958.
Hogben, Lancelot. *Mathematics for the Million*. 4th edn, London, 1967 and repr.
Murray, H. J. R. *A History of Board Games other than Chess*. Oxford, 1952.
Neugebauer, O. *A History of Ancient Mathematical Astronomy*. 3 vols. Berlin/Heidelberg/New York, 1975.
Pedersen, O. and Pihl, M. *Early Physics and Astronomy*. New York and London, 1974.
Rogers, J. T. *The Story of Mathematics*. 2nd edn, London, 1979.
Sarton, George. *A History of Science*, Vols 1–2. London and Cambridge, Mass., 1953–9.
Singer, Charles *et al.* (eds). *A History of Technology*, Vols 1–2. Oxford, 1954–6 and repr.
Skinner, F. G. *Weights and Measures: their Ancient Origins and their Development in Great Britain up to AD 1855*. London, 1967.
Smith, David E. *History of Mathematics*. 2 vols. Boston, 1923.
Taton, René. *Ancient and Medieval Science from Prehistory to AD 1450*, transl. A. J. Pomerans. London, 1963.
Waerden, B. L. van der. *Science Awakening*, transl. A. Dresden. 3rd edn, New York, 1971.

Early Period

Aaboe, A. *Episodes from the Early History of Mathematics*. New York, 1964.
Bruins, Evert M. 'Egyptian arithmetic', *Janus* 68 (1981), 277–318.
Chace, A. B. *et al. The Rhind Mathematical Papyrus*. 2 vols. Mathematical Association of America, Oberlin, Ohio, 1927–9, repr. 1979.
Edwards, I. E. S. *The Pyramids of Ancient Egypt*. Harmondsworth, 1961 and repr.
Gardiner, Sir Alan. *Egyptian Grammar*. 3rd edn, London, 1957.
Gillings, R. J. *Mathematics in the Time of the Pharaohs*. Cambridge, Mass., 1972.
Nemet-Nejet, Karin. *Late Babylonian Field Plans in the British Museum*. Studia Pohl, ser. maior 11. Rome, 1982.
Neugebauer, O. *The Exact Sciences in Antiquity*. 2nd edn, Providence, R. I., 1957 and New York, 1962.
Neugebauer, O. and Parker, R. A. *Egyptian Astronomical Texts*. 3 vols. London, 1960–69.
Neugebauer, O. and Sachs, A. J. *Mathematical Cuneiform Texts*. American Oriental Series, vol. 29. New Haven, Conn., 1945.
Parker, R. A. *The Calendars of Ancient Egypt*. Chicago, 1960.
Peet, T. Eric. *The Rhind Mathematical Papyrus*. London, 1923.
Robins, Gay and Shute, Charles. *The Rhind Mathematical Papyrus*. London, 1987.
Sachs, A. J. 'Two neo-Babylonian metrological tables from Nippur', *Journal of Cuneiform Studies* 1 (1947), 67–71.
Sachs, A. J. 'Babylonian mathematical texts', *Journal of Cuneiform Studies* 1 (1947), 219–40, and 6 (1952), 151–6.
Stephenson, F. R. and Walker, C. B. F. *Halley's Comet in History*. London, 1985.
Struik, D. J. 'Minoan and Mycenaean numerals', *Historia Mathematica* 9 (1982), 54–8.
Wheeler, N. F. 'Pyramids and their purpose', *Antiquity* 9 (1935), 5–21, 161–89, 292–304.

Greece and Rome

Allman, G. J. *Greek Geometry from Thales to Euclid*. Dublin and London, 1888; repr. New York, 1976.
Archimedes. *Works*, transl. T. L. Heath. Cambridge, 1897.
Austin, R. G. 'Roman Board Games', *Greece and Rome*, 1st ser., 4 (1934–5), 24–35, 76–82.

Boëthius, Axel and Ward-Perkins, J. B. *Etruscan and Roman Architecture*. Harmondsworth, 1970.

Brumbaugh, R. S. *Ancient Greek Gadgets and Machines*. New York, 1966; repr. Westport, Conn. 1975.

Camp, J. M. *The Athenian Agora*. London, 1986.

Columella, *De re rustica*, ed. H. B. Ash *et al.* 3 vols. Loeb Classical Library, London and Cambridge, Mass., 1941–55 and repr.

Detlefsen, M. *et al.* 'Computation with Roman numerals', *Archives for History of Exact Sciences* 15 (1976), 141–8.

Dijksterhuis, E. J. *Archimedes*. Copenhagen, 1956; New York, 1957.

Dilke, O. A. W. *Greek and Roman Maps*. London, 1985.

Dilke, O. A. W. *The Roman Land Surveyors: an Introduction to the Agrimensores*. Newton Abbot, 1971.

Dilke, O. A. W. *Surveying the Roman Way* (pack of instructions). School of Classics, University of Leeds, 1980.

Euclid. *The Thirteen Books of Euclid's Elements*, ed. T. L. Heath. 3 vols. Cambridge, 1908.

Gibbs, Sharon L. *Greek and Roman Sundials*. New Haven, Conn., and London, 1976.

Goodspeed, E. J. 'The Ayer Papyrus: a mathematical fragment', *American Journal of Philology* 19 (1898), 25–39.

Haselberger, L. 'The Construction Plans for the Temple of Apollo at Didyma', *Scientific American*, December 1985, 126–32.

Heath, T. L. *A History of Greek Mathematics*. Oxford, 1921, repr. 1965.

Hultsch, F. *Griechische und römische Metrologie*. 2nd edn, Berlin, 1882.

Hultsch, F. (ed.) *Reliquiae scriptorum metrologicorum*. Teubner series, Leipzig, 1864–6 and repr.

Huxley, G. L. *Anthemius of Tralles: a Study in Later Greek Geometry*. Greek, Roman and Byzantine monographs 1. Cambridge, Mass., 1979.

Kent, J. P. C. *Roman Coins*. London, 1978.

Knorr, W. R. *Ancient Tradition of Geometric Problems*. Basle/Stuttgart/Boston, 1986.

Kraay, C. M. *Greek Coins*. London, 1966.

Landels, J. G. *Engineering in the Ancient World*. London, 1978.

Lang, Mabel and Crosby, Margaret. *The Athenian Agora*. Vol. X. *Weights, Measures and Tokens*. Princeton, N.J., 1964.

Lloyd, G. E. R. *Early Greek Science: Thales to Aristotle*. London and New York, 1970.

Lloyd, G. E. R. *Greek Science after Aristotle*. London and New York, 1973.

Marrou, H. I. *A History of Education in Antiquity*, transl. George Lamb. London and New York, 1956.

Mueller, I. *Coping with Mathematics (the Greek Way)*. Booklet, Morris Fischbein Center, Chicago, 1980.

Price, D. J. de Solla. 'Gears from the Greeks: the Antikythera Mechanism', *Transactions of the American Philosophical Society* n.s. 64, pt. 7 (1974).

Price, D. J. de Solla. 'The water clock in the Tower of the Winds', *American Journal of Archaeology* 72 (1968), 345–55.

Ptolemy. *Almagest*, transl. G. J. Toomer. London, 1984.

Richardson, William F. *Numbering and Measuring in the Classical World*. Auckland, 1985.

Robertson, D. S. *A Handbook of Greek and Roman Architecture*. 2nd edn, new impr., Cambridge, 1969.

Sandys, Sir John E. (ed.). *A Companion to Latin Studies*. 3rd edn, Cambridge, 1921 and repr.

Stahl, William H. *Roman Science: Origins, Development and Influence to the later Middle Ages*. Madison, Wisc., 1962.

Taisbak, C. M. 'Roman numerals and the abacus', *Classica et Mediaevalia* 26 (1965), 153–60.

Thomas, Ivor (ed. and transl.). *Selections illustrating the History of Greek Mathematics*. 2 vols. Loeb Classical Library, London and Cambridge, Mass., 1939–41.

Thomson, J. Oliver. *History of Ancient Geography*. Cambridge, 1948.

Tod, Marcus Niebuhr. *Ancient Greek Numerical Systems* (reprints of six articles in *Annual of the British School at Athens* and *Journal of Hellenic Studies*). Chicago, 1979.

Turner, J. Hilton. 'Roman elementary mathematics: the operations', *Classical Journal* 47 (1951), 63–74, 106–8.

Vitruvius. *De architectura*, ed. F. Granger. 2 vols. Loeb Classical Library. London and Cambridge, Mass., 1931–4 and repr.

Warren, J. *Greek Mathematics and the Architects to Justinian*. (Art and archaeology research papers.) Horsham, 1976.

Whibley, L. *A Companion to Greek Studies*. 4th edn, Cambridge, 1971.

White, K. D. *Greek and Roman Technology*. London, 1984.

Index